古墅天成

ODE TO MASTERPIECES

APPRECIATION OF CLASSICAL EUROPEAN VILLAS

深圳视界文化传播有限公司 编

欧式古典别墅赏析

中国林业出版社
China Forestry Publishing House

图书在版编目（CIP）数据

古墅天成：欧式古典别墅赏析 / 深圳视界文化传播有限公司编． -- 北京：中国林业出版社，2017.1
ISBN 978-7-5038-8917-2

Ⅰ．①古… Ⅱ．①深… Ⅲ．①别墅－建筑设计－作品集－世界－现代 Ⅳ．①TU241.1

中国版本图书馆CIP数据核字（2017）第023437号

编委会成员名单
策划制作：深圳视界文化传播有限公司（www.dvip-sz.com）
总 策 划：万绍东
编　　辑：杨珍琼
装帧设计：潘如清
联系电话：0755-82834960

中国林业出版社 · 建筑分社
策　　划：纪　亮
责任编辑：纪　亮　王思源

出版：中国林业出版社
（100009 北京西城区德内大街刘海胡同 7 号）
http://lycb.forestry.gov.cn/
电话：（010）8314 3518
发行：中国林业出版社
印刷：深圳市雅仕达印务有限公司
版次：2017 年 3 月第 1 版
印次：2017 年 3 月第 1 次
开本：215mm×275mm，1/16
印张：18
字数：300 千字
定价：280.00 元 (USD 48.00)

PREFACE
序言

欧式风格于17世纪盛行欧洲，继承了巴洛克风格中豪华、动感、多变的视觉效果，也吸取了洛可可风格中唯美、律动的细节处理元素，以豪华、大气、奢侈为突出点，色彩华丽。它在形式上以浪漫主义为基础，材质上常使用大理石、多彩的织物、精美的地毯，精致的法国壁挂，风格整体豪华、富丽，充满强烈的动感效果，受到上层人士的青睐。

The European style prevailed in 17th century in Europe, inherited the luxurious, dynamic and changeable visual effects of baroque style, and absorbed the aesthetic and rhythmic detail treatments of rococo style with luxury, magnificence and extravagance as the highlights, which is colorful and gorgeous. It is based on romanticism in form. It usually adopts marbles, colorful fabrics, exquisite carpets and delicate French wall hangings. The overall style is luxurious and magnificent full of strong dynamic effects, which is loved by the upper class.

欧式风格以精美的造型、华丽的装饰、浓烈的色彩来达到雍容华贵的装饰效果。造型上，以罗马柱、欧式拱门、壁炉等经典造型突出空间上的富丽堂皇（图1）。装饰上，选用华丽名贵的家具、家饰等装点整体空间，散发着古典优雅的贵族气质（图2）。色彩上，偏于描金绘银的闪亮色系，整体用色偏厚重沉稳，呈豪华奢侈的视觉形态（图3）。

（图1）

The European style shows elegant and gorgeous decorative effects by exquisite modeling, magnificent furnishings and strong colors. The classic modeling, such as Roman column, European arches and fireplaces highlight the magnificence and gorgeousness of the space as shown in the picture one. The decorations use gorgeous and rare furniture and furnishings to decorate the whole space, which manifests classical and elegant noble temperaments as shown in picture two. The colors prefer shining tones such as gold and silver. The whole colors are dignified and sedate, which presents luxurious and sumptuous visual forms as shown in picture three.

（图2）

《古墅天成——欧式古典别墅赏析》选用的是原汁原味的来自欧洲领先设计师最新设计的欧式古典别墅，很好地呈现了欧式风格设计传统的历史痕迹和浑厚的文化底蕴。通过完美的案例展示，精益求精的细节说明，带给读者以最高的赏阅境界。

Ode to Masterpieces—Appreciation of Classical European Villas chooses the latest classical European villas designed by original leading designers in Europe, which presents the traditional historic traces and heavy cultural connotations of designs in European style. The perfect project displays and excelsior detail explanation can bring the readers the highest state of enjoyment.

（图3）

CONTENTS 目录

006	豪宅艺术 THE ART OF MANSION	154	绝佳观景，山顶美宅 A HILLTOP RESIDENCE WITH PERFECT VIEWS
020	艺术的交织 THE INTERWEAVING ART	170	古雅世界的魅惑 OLD WORLD CHARM
052	灵感庄园 INSPIRING MANOR	188	迷情帝国之家 CHARMING IMPERIAL HOME
076	历史文化的呼唤 THE CALL OF THE HISTORIC CULTURE	204	文艺复兴之家 RENAISSANCE HOME
086	古雅城堡 A CLASSICAL AND ELEGANT CASTLE	212	金色之家 GOLD HOME
096	异域风情 THE EXOTIC STYLE	234	绮思幽梦 PECULIAR THOUGHT AND QUIET DREAM
114	传承法式经典 INHERITING FRENCH CLASSIC	254	艺术家 AN ARTISTIC HOME
128	优雅品质生活 ELEGANT QUALITY LIFE	260	尊贵之家 AN EXALTED HOME
140	优山大墅 MOUNTAIN VILLA	280	犹享宫廷美墅 ENJOYING THE PALACE VILLA

THE ART OF MANSION
豪宅艺术

DESIGN CONCEPT | 设计理念

According to the owner's requirements, the design concept is mainly oriented to keep and adapt the existing, rich and "With Grandeur" ambience to a very chic, cozy and comfortable family house. The mix-and-match of different styles makes the house unique. The improved baroque style reduces complicated and heavy decorations, but keeps the elegant, beautiful, magnificent and majestic temperaments. The rococo slimness and softness contrast and supplement with the baroque vigorousness in some degree. The modern furniture and colors are concise and fashionable, which is more suitable for modern family living. The screen in Chinese style is the highlight of the European villa. The classic baroque and rococo styles collocate with the bright and light modern style, which integrates life philosophy with art philosophy harmoniously and reflects consummate design foundations of the designer.

设计公司/设计师/摄影师/地点/面积

Design company : Viterbo Interior Design
Location : Lisbon, Portugal
Designer : Gracinha Viterbo
Area : 2,340 m²
Photographer : Frederic Ducout

设计师根据屋主的要求呈现出的设计理念主要是保持现有的浓郁"宏伟"氛围，并且让这种氛围适应于一个别致、温馨而舒适的家庭。不同风格的混合搭配让房屋独具特色，经过改良的巴洛克式风格减少了繁复厚重的装饰，但依旧保留了雅致优美和大方庄严之感。洛可可式的纤细柔美，在一定程度上与巴洛克式风格的浑厚形成对比又相互补充。现代风格的家具和配色简约时尚，更适合现代家庭的居住，中式的屏风是整个欧式别墅中的亮眼之处。经典的巴洛克式风格和洛可可式风格，搭配明丽轻快的现代风格，将生活哲学和艺术哲学和谐融为一体，体现出设计师醇熟的设计功底。

餐厅：

以三层水晶吊灯为中心的餐厅能满足十二人的就餐需求，中黄黑的餐桌与木质地砖颜色一致，餐桌椅上金色装饰与天花相呼应，白色墙体雕饰又和天花互为一体，草绿色窗帘和文艺复兴风格壁画外的雕饰迎面成趣，整个空间呈现出的清新明丽让每天的就餐氛围更加愉悦。

客厅：

多窗设计让整个客厅空间更显大气明亮，金色、米黄色和灰色搭配让空间保持风格统一。巴洛克式风格镶金天花雕饰独具特色，椭圆形的文艺复兴时期的壁画，古典、生动又细腻。竖琴设计彰显出屋主的音乐品味，灰汁色窗帘和落地灯的设计是现代风格，客厅是古典与现代风格融合的经典之作。

SPACE PLANNING | 空间规划

这座别墅因其独特的设计和历史背景成为一个具有象征性的建筑，经过改造、室内设计和装饰之后共有三层。主层为社交区域和主人套房，所有空间都有着独特的装饰、画作、帝家丽手绘墙纸和金银饰品，此外还设有私人办公室、图书室和冬日花园，动线走廊上还铺有错视画的木质地板和专门的手工缝制葡萄牙地毯，墙壁上则镶有威尼斯艺术风格的奥尔西尼马赛克手工雕刻嵌板。一楼的家庭活动室有着更为现代和轻松的设计，可以观赏到壮观的塔霍河及周边的全景。客卧的设计则比较古典，但在纺织物和墙纸上采用了现代的配色方案。地下层为服务区、酒窖、水疗中心、室内泳池、健身房、按摩室和休闲室，别墅空间规划合理、灵动、实用，给屋主带来温馨舒适的居住体验。

书房：

　　书房以过道门和窗为中心设计隔层书架，上面摆满的书籍彰显出屋主的文化内涵。壁画和黑色摆件与其他空间在风格上保持关联性，银白色和深棕色的沙发是古典与现代风格的均衡搭配，红色抱枕和窗帘为书房增添一丝饱满热情，休闲时光来此看看书和小憩一会，也是生活难得的闲情逸趣。

端景：

  定制的大理石桌，桌面和地砖、墙面使用一致的天然大理石材质，绿色和古铜色纹理流淌交融在一起，自然天成，凸显出法式古典风格的宫廷贵族气息。古铜色的桌脚有精美的曲线雕饰，展示出巴洛克风格独具特色的富丽与典雅。墙面上挂置的中世纪油画，展现出意大利文艺复兴时自由民主思想。

THE INTERWEAVING ART
艺术的交织

DESIGN CONCEPT | 设计理念

The house is designed to meet the owner's emotional belonging to home and promote the popularity of artistic style from the basic sensory experiences. The owner is found of Gothic style, so the challenge presented to the designers is to remodel the existing interior and reconcile the Mediterranean and Italianate exterior with their love of everything Gothic to produce artistic beauty. The design team spends months researching Gothic revival and decides to use Gothic and Italianate styles in the home. Gothic revival is an intimidating style whose main tone is solemn, classical and sacred. The whole is a little gloomy and dark, but is also luxurious and elegant. The designers use exquisite lines, bold space plans and reasonable applications of lights to create a unique, luxurious, charming and multi-decorative mysterious artistic state.

设计公司/设计师/摄影师/地点/面积

Design company : The Design Firm
Photographer : Julie Soefer
Designers : Kara Wuellner, Jennifer Meeks, Rachel Gracriamaria, Crystal Tennant, Mary Hare
Location : Houston, Texas, USA
Area : 1207 m²

设计师的设计初衷是为了满足屋主对家的情感归属，同时也可以从最基础的感官体验上推动艺术风格的普及。屋主偏爱哥特式风格，而设计师所面临的挑战则是改造原有的室内设计，使原有的室外建筑设计风格与业主偏好的风格相协调，并产生艺术美的效应。设计团队耗时良久来研究哥特式复兴风格，最终决定在室内运用哥特式和意大利两种风格。哥特式复兴风格是一种令人敬畏的风格，主要格调庄严、古典、神圣，整体偏沉郁黑暗却不失华丽优雅。设计师用考究的线条，大胆的空间设计，光线的合理运用，打造出一个奇特的、奢魅的、装饰多元化的神秘艺境。

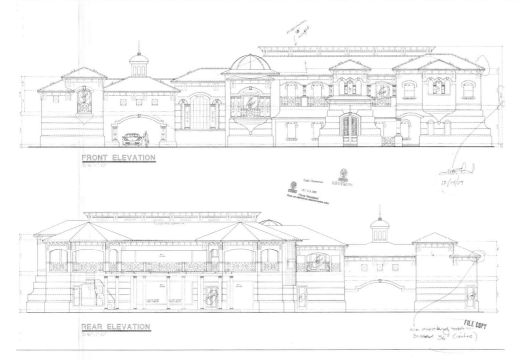

客厅：

在怀旧复古的昏黄色调中，深红色、黑色和黄褐色的穿插与交织，共同创造出了一个优雅的、引人注目的、更具历史传统氛围的别墅空间。挑高的设置，精致的铁艺，在浑厚的格调中领略优雅的线条美。层层递进的弧形拱门造型背景墙，大气的同时提升了空间感。而独具历史感与民族特色交织的帘幔设计又打破了背景墙的单调，更显艺术性。

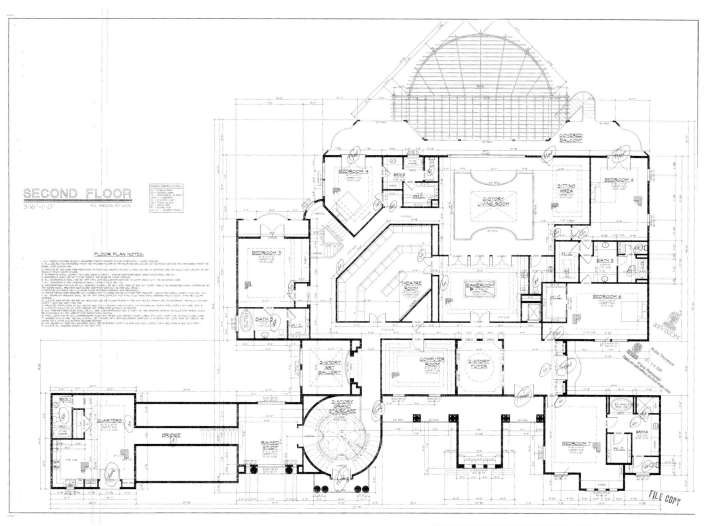

SPACE PLANNING | 空间规划

设计师是在别墅原有的户型、面积、走向、结构等空间基础上进行再创造。一层按功能性大概划分为客厅、餐厅、厨房等区域。一层的空间功能更适合开放式或半开放式的空间模式，因此餐厅与客厅之间采用开放式的空间格局，视野更广，活动空间更加宽敞明亮，同时家人互动的氛围也更为舒适。二楼的区域划分有起居室、阅读区、卧室、品酒区、影音室、娱乐区等，设备齐全。

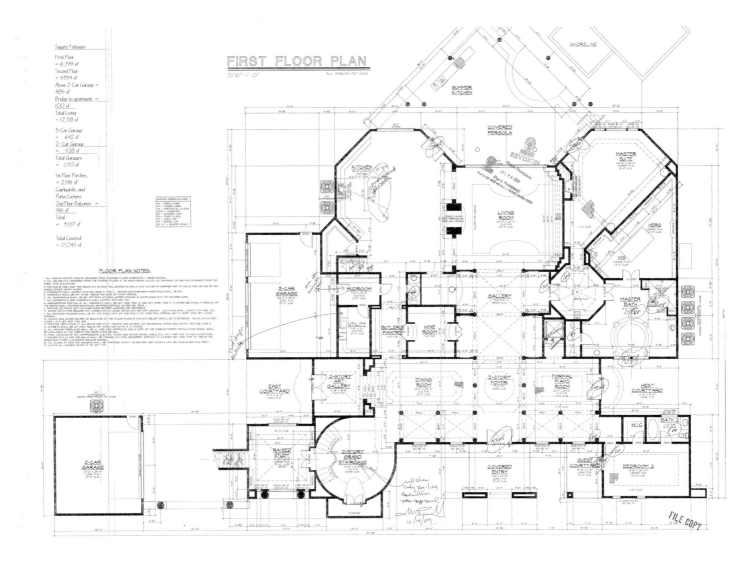

阅读空间：

二楼阅读区的整个空间调性基本涵盖了这栋别墅的软装基调。墙面用天然文化原石铺陈，不规则的拼接加上石材的自然肌理，粗犷奔放的风格体现了空间原始的唯美与宁静，更强调了生命的本质与内涵，这种墙面的装饰手法在欧式与美式古典设计风格中尤为常见。红色、金色、黑色构成了这个空间的色彩组合，这是欧式古典风格中最纯正的色彩搭配，能够呈现出空间特有的贵族气息。红金相间、厚重感极强的欧式窗帘与椅子上的一袭红毯是空间的提神之笔，而两张庄重气派、精雕细刻、雍容优雅的新古典风格椅子则是安详、沉静的智者，让空间色调沉稳和谐。圆形的吊顶，欧式卷叶纹的复刻以及涡旋的层次感，让空间更加精致奢雅，中间焚烧画卷的抽象图绘传达一种欲说还休的神秘意境。

玄关之美：

　　玄关是别墅出入的必经之地，也是设计师整体设计思想的浓缩，它在房间装饰中起到点睛的作用。欧式别墅的入室玄关并不像中式玄关那样需要隐蔽，反而更多的是敞亮通透。一楼玄关处挑高的中空，垂落的水晶吊灯，带给入室者恢弘大气的直视感。玄关中间放置的圆形供桌，精致的雕花、复古的铁艺、大理石与实木的夹合式设计更显古朴与尊贵，加上高挑的狮脚蜡台与昏黄的烛火，为这个空间增加了颇多的神秘色彩。过廊采用连续的拱壁与波斯灰罗马柱相承接的方式，整体给人一种线条的韵律感、量感、空间感和丰富而有变化的立体感。墙壁在色调统一的基础之上讲究壁面的形式美，设计师用乳胶漆做旧的手法，呈现出一种斑驳的、复古的沧桑之美。

餐厅：

　　设计师充分利用空间环境、艺术美学、个性偏好等多种复杂元素的创造性融合，家具、装饰画、花艺、布艺、灯饰等软装与硬装巧妙协搭。无论是深色的硬装墙面与米色的布艺沙发，还是精细考究的吊顶与浪漫温馨的吊灯，都在视觉上将数学与美学完美结合，散发出迷人的空间魅力。

主浴室：

主浴室是哥特式风格的缩影，水疗浴缸的顶部采用尖肋拱顶造型，四根罗马柱作为重力支撑，木质的拱顶和拱底相呼应，因波斯灰罗马柱的介入而不显沉闷。设计师保留了意大利风格的拱形窗户，采用进口的彩色窗格玻璃镶嵌，表达着对哥特式巅峰之作兰斯大教堂的由衷敬意。

次卧：

次卧之一选择的是地中海风格，这种风格最大的魅力来自其纯美的色彩组合，这里运用了地中海风格最典型的颜色搭配：西班牙蔚蓝色的海岸与白色沙滩之蓝白搭配。海与天的明亮色彩、仿佛被水冲刷过后的白墙，自然、大胆。白色的窗幔，质地柔软，营造浪漫清新的氛围；蓝白相间的条纹元素也是地中海风格里常用到的，而这里的条纹地毯蓝白之间加了黄、绿色，延续了椅子的色调，色彩层次也更为丰富。白色珊瑚礁形态的台灯，天然海螺的小饰物，与地毯同样色彩基调的装饰画，角落里的枯枝盆景，这些细节的处理打破了空间色彩的单一乏味性，让空间内容更加富有人文精神和艺术气质。

INSPIRING MANOR
灵感庄园

DESIGN CONCEPT | 设计理念

This manor is designed to be the perfect expression of the clients – who they are and how they live. The clients envision a Gothic house that focuses on authentic details and quality both inside and outside and that is neither typical of the area or the times. So it's a big challenge to the designer. The foundations of design draw primarily from a combination of Gothic, Tudor and Jacobean styles and periods and regions throughout Europe. The designer adheres to a philosophy of using real materials, so there is nothing "faux" in the house. And even though the majority of materials the designer uses are new, he builds and finishes them with old-world techniques including mortise, tenon joinery and a seven-step process to the wood finish, the final step being hand-waxing. Finally the house is designed to be a British manor that can withstand the baptism of the years like other historic British manor.

设计公司/设计师/摄影师/地点/面积

Design company : Cravotta Interiors
Location : Austin, TX

Designer : Mark Cravotta
Area : 1,300 m²

Photographer : Paul Bardagjy

　　这个庄园的设计是对客户的完美体现，他们是什么样的人，他们又是怎样生活的。客户设想了一座哥特式的房子，要求室内室外都注重原汁原味的细节和质量，却没有典型的区域或时代要求，因此，这对设计师来说是个极大的挑战。设计的基础主要是从欧洲的哥特式、都铎式和詹姆士等风格的不同时期和地区的结合中获得灵感。设计师坚持使用纯自然的材料，整个房子中几乎没有任何人造材料。尽管使用的大部分材料都是新的，但设计师建造这座房子使用的却是古老的技艺，包括榫接、榫眼细木工业以及木工的七个步骤，最后一步就是手工打蜡。最终这座房子被打造成一座英式庄园，像那些具有重大历史意义的英国庄园一样，能经得起岁月的洗礼。

客厅：

　　这座英式庄园的主要特点是造型繁琐，有很浓厚的教堂气息，给人一种庄重、神秘、严肃的感觉，客厅气势尤其突出。为了更好地配合空间的调性，软装搭配也秉承了这些特点，沉稳的格调中透露着鲜活气息，丰满与复古相结合，凸显出整个空间的大气恢弘。

　　纯天然的石壁打造与雕刻石像构成空间的整个壁像，通透的圆形拱门加上高镂空的客厅，霸气的即视感。同时金色的烛台大吊灯，一泻而下，皇家范十足。顶面全实木的雕刻，做工精细，彰显出贵族气质的优雅与洒脱。

LIVING ROOM / DINING ROOM

家庭室：

由于空间中大量使用实木雕刻，因此木色是整个空间的颜色基调。深红印花沙发、紫色印花地毯、还有长长的银灰色帘幔互相依偎，典雅中透露出细腻的生活气息。

SPACE PLANNING ｜ 空间规划

业主讲究生活的质量与格调，设计师在建筑原有的空间结构基础之上，根据业主的需求重新进行规划设计。整个住宅根据功能性分区，全面精细。客厅、餐厅、厨房、家庭休闲室构成整个空间的动线，增强了室内的开放性，便于家人之间的情感交流。

休闲吧：

　　格栅实木吊顶，全实木打造的吧台和橱柜，拼接实木地板，给整个休闲吧营造了一种精致、沉稳的氛围，而淡雅印花的躺椅则增添了几分生活情趣。休闲吧延续整个空间的格调，从椅子、吧台立面、橱柜表面到墙面，都是纯手工雕刻和打蜡，意在为业主打造一个精美的休闲空间，也很好地体现了设计师对设计的精益求精。

BILLIARD ROOM

065

台球室：

　　木格吊顶厚重大气，水晶吊灯温馨舒适，实木书橱古色古香，雅致印花地毯灵动活波，红色球桌静静置于空间的中心，在这里，既可以畅游书海，荡涤心灵，又可以休闲娱乐，强身健体。可以说，二者无缝的结合是设计师的匠心巧思。更值得称赞的是拱形木格玻璃落地窗的设计，大面积增加了室内的采光和通风，偶尔的抬头间，窗外亦是一处可以欣赏的风景。

主卧：

温馨淡雅的色系组合，精致繁复的天花雕刻，构成整个空间的格调，雅致中显贵气，细腻中品柔情。

家庭影音室：

　　铜制大门，做工精细，两头威武的雄狮雕刻左右，给人一种雄厚感。推开门，深红的沙发、地毯、幕帘，延续了庄园大气热烈的氛围。长方形的空间设计，地毯的铺设以及布艺沙发的使用，起到了吸声和隔声的效果。幕帘装饰花纹、柱头雕刻、吊顶线条雕刻从细节处凸显了设计师的鬼斧神工。整个家庭影音室给人一种殿堂级的视觉盛宴，身与心的放松享受。

WINE CELLAR

酒窖：

设计师希望将酒窖打造成与庄园里其他部分一样，看起来有悠久的历史但却依然保存完好的效果。整个酒窖全部由石头打造，并配置有全手工的橱柜和木制品。墙壁和外墙石头一样，融合了四个品种的石灰岩和砂岩，地面则是仿古胡桃木纹理的石灰石地板，天花板是陶瓷瓷砖。木制品和橱柜都是美国黑桃木材质，表面却处理得如同法国桃木。长凳和19世纪50年代梦幻般的木壁炉，都是从法国和比利时边境的制造商买来的，桌椅都是法国桃木材质，均为私人定制。藏酒于如此精心制造的酒窖中，想必酒也会更香醇可口。

THE CALL OF THE HISTORIC CULTURE
历史文化的呼唤

DESIGN CONCEPT | 设计理念

The largest estate on Marco Island, GREYSTONE MANOR, is blessed with a unique style like no other on the southwest coast of Florida. The owner has spent over two and a half years to renovate the manor in order to seek for an ideal residence. In the premise of following the owner's preferences, the designer uses her personal aesthetic experience to plan the three acres land and makes it a better manor both outside and inside. The manor with traditional European style exterior is located in a wide estate, which presents a magnificent and majestic sense. There are over one hundred and fifty specimen of tropical trees and plants with unexpected statuaries from Europe, which leaves more tranquil and private spaces for the manor. After the designer's repeated researches, the interior adopts Roman style in order to better fit the exterior environment. The reasonable applications of space, lines, furnishings and lights in the house deduce the dense historic sense and cultural flavor of Roman style incisively and vividly.

设计公司/设计师/摄影师/地点

Design company : Ariam Interiors
Location : Florida, USA
Designer : Maria Coords
Photographers: Maria Coords, Ariam Interiors

　　灰石庄园作为马可岛上最大的地产，与佛罗里达州西南海岸上的风格迥然不同。屋主花费两年半的时间翻新这座庄园，只为寻找心中的理想居所。设计师在遵循屋主喜好的前提下，同时调动自身审美经验，为屋主规划了这片占地三亩，内外相得益彰的庄园。屋外传统欧式风格的建筑坐落在开阔的庄园中，大气庄严之感顿然而生。150种热带绿植围绕生长，随处可见的欧洲雕像，又给这片庄园增添许多宁静和隐私的空间。经过设计师的反复研究，与室外环境更好地贴合，室内最终采用罗马式风格，屋内空间、线条、装饰和灯光的合理运用，将罗马式风格所具有浓厚的历史感和文化气息演绎得淋漓尽致。

入门长廊：

　　入门长廊是屋主出入的必经之地，是室外和室内视觉过渡的重要场所，它也是屋主和设计师室内审美体现的首要之地。长廊设计较为狭长，给屋主一个由室外到室内的视野过渡。绿植相迎，古铜材质做旧的仿古吊灯，昏黄色调的灯光照应拱形门柱，穿越长廊感受到扑面而来的古典气息；奶白色的墙面包裹着轻金属色的壁面铁艺装饰，凹凸有序的罗马柱，使空间更富有立体艺术感；米黄色的大理石地砖上搭配黑色和深黄色框形纹理，古朴庄严又不失尊贵气息。

SPACE PLANNING | 空间规划

　　合理的空间布局更能让屋主体验到家所赐予的人文关怀。设计师在庄园原有的空间结构上进行再创造，一层按功能性大概分为玄关、入门长廊、客厅、厨房、餐厅等区域。客厅上方采用高耸的圆拱形屋顶，使空间向上延伸，更显宽敞大气，大空间的设置使聚会沙龙随意进行；开放式的厨房与餐厅密切相连，让进餐体验更加舒适；休息室的软榻沙发，立刻舒缓身体的疲劳。一层空间动静结合，让回家成为一种享受。二层区域主要划分为阅读室、卧室、起居室等，深色系的阅读室营造出浓郁的书香气息，更适合屋主安静地阅读。奶白色和米黄色为主调的卧室，舒适大方，有助于屋主入梦。

厨房餐厅：

厨房餐厅作为一家人共进晚餐的场所，它也承接着一家人情感交流的重要职责。舒适开阔的厨房和餐厅密切相连，整体空间采用奶白色调，更显轻松舒适。树脂材质的欧式仿云石吊灯投射出淡黄的灯光，房顶筒灯和暗藏灯带的设计，既满足了空间照明需求，又营造出舒适温馨的进餐氛围。品类丰富的厨餐用具，不间断的欢声笑语，一日三餐的精致饮食和情感交流即刻从这里出发。

阅读室：

　　阅读不仅是一种获得知识的渠道，更是一种沉淀内心、修养身心的方式。具有书香气息的红褐色赋予空间更多历史文化内涵，也是本案的点睛之笔。阅读室墙面设有壁炉，不仅可供冬日取暖，也能干燥室内环境。木质书柜和书桌相互照应，营造出一种安宁的阅读环境。阅读室中的装饰品丰富多元，野兽雕、仿古吊钟、茶具、插花等，打破时空地域的限制，将古典现代风格有序搬进书房，彰显出一种厚重的人文底蕴和文化气息。

主卧：

卧室不仅是屋主身体的放松场所，也是精神的休憩之地。主卧硬装设计采用欧式风格，奶白色的墙面和咖啡色高挑的圆拱形屋顶，散发出舒适典雅的贵族气息。软装设计与硬装相呼应：牛奶白和米黄色的床品，古朴的蜡烛式水晶吊灯和刻意做旧的床头漆艺台灯，浅咖色的欧式沙发与细花图案的窗帘帷幔，深咖色毛绒地毯等，与屋顶、墙面颜色相呼应，系统而完整的设计手法，使主卧空间有序地融为一体。

A CLASSICAL AND ELEGANT CASTLE
古雅城堡

DESIGN CONCEPT | 设计理念

This lavish newly built Mediterranean style waterfront mansion is located on West Isle Place in The Woodlands, Texas, with advantaged beautiful natural scenery. So beauty, luxury and uniqueness are all top priorities throughout the construction period, which is comparable with the surrounding scenery. Thick layers of crown molding, marble staircases, Venetian plaster walls, and 24 carat gold hardware barely scratch the surface of the unbelievable features in this home. The abundance of plants planted in pots greatly fits in a chic setting. A lush green looks great on a white background of the original house. The team of architects and designers do their best to give the owners an elegant home that is cheering and surprising on the outside and inside, looking at the amazing modern castle, involuntarily feeling like in a fairy tale.

设计公司/建筑师/地点

Design company : Gary Keith Jackson Design, Inc.
Location : The Woodlands, TX, USA
Architect : Gary Jackson

FIRST FLOOR PLAN 一层平面图

1. WET SAUNA	17. BTLR/WET BAR	1、湿蒸室	17、管家房/酒吧
2. DRY SAUNA	18. WINE GROTTO	2、干蒸室	18、酒窖
3. SPA ROOM	19. 2-STORY DINNING	3、水疗室	19、两层高餐厅
4. SUMMER KITCHEN	20. FRONT DRIVE	4、夏季厨房	20、前车道
5. BATH	21. ENTRY PORCH	5、浴室	21、入口玄关
6. COVERED VERANDA	22. 2-STORY FOYER	6、带顶走廊	22、两层高门厅
7. THEATER ROOM	23. 2-STORY LIVING ROOM	7、影音室	23、两层高客厅
8. FAMILY ROOM	24. POOL RESERVE	8、家庭活动室	24、泳池储备区
9. UTILITY	25. SITTING ROOM	9、杂物间	25、起居室
10. POWDER	26. MASTER BEDROOM	10、化妆室	26、主卧
11. CLOSET	27. LIBRARY	11、壁橱	27、图书室
12. SIDE ENTRY	28. HIS	12、侧面入口	28、男卫
13. TWO CAR GARAGE	29. POWDER/BATH	13、双车车库	29、化妆间/浴室
14. KITCHEN	30. WET BAR	14、厨房	30、酒吧
15. BREAKFAST	31. MASTER BATH	15、早餐区	31、主卫
16. PANTRY	32. HERS	16、餐具室	32、女卫

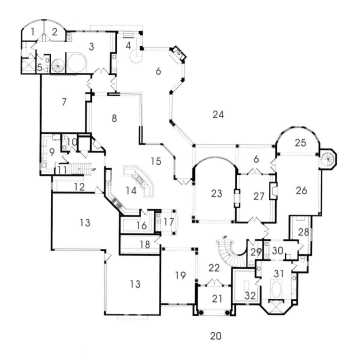

SECOND FLOOR PLAN 二层平面图

1. EXERCISE ROOM	8. POWDER	1、健身房	8、化妆室
2. BEDROOM	9. GAME ROOM	2、卧室	9、游戏室
3. BATH	10. BALCONY OVERLOOK	3、浴室	10、俯瞰阳台
4. CLOSET	11. OPEN TO BELOW	4、壁橱	11、上空
5. COVERED BALCONY	12. RETREAT	5、带顶阳台	12、休息室
6. SITTING ROOM	13. ATTIC STORAGE	6、起居室	13、阁楼储藏室
7. ELEVATOR		7、电梯	

　　这座新建的奢华地中海风格的海滨别墅坐落于德克萨斯州的伍德兰市，拥有得天独厚的美丽的自然景观，因此，设计师在设计时优先考虑的就是美观、豪华以及独特性，使其能与周围的美景媲美。极富层次的顶冠造型、大理石楼梯、威尼斯灰泥墙和24克拉黄金器具成为了这座房子的代表性特征。大量的盆栽植物与别致的环境相融合，郁郁葱葱的绿色在这座白色调房屋的映衬下显得非常美丽。整个建筑和设计团队都在尽心尽力为业主打造一个由内而外令人惊叹的优雅家园，看着这个迷人的现代城堡，有一种身处童话故事中的感觉。

客厅：

　　为了装饰白色墙壁，设计师使用了威尼斯灰泥和天然大理石，这些华丽的石材通常被用于宫殿设计。吊灯由水晶和黄金打造，设计非常奢华美观。此外，大钢琴的加入为整个银灰色调的空间增添了一抹艺术气息。

厨房：

　　设计师想要在具有宫廷风格的住宅中打造一个精细的厨房，环抱式的设计，精致的门后面隐藏着一个大型冰箱，白色的橱柜内有各种嵌入式的电器。地板上铺设的西班牙米黄大理石光滑细腻，体现了业主对高品质生活的追求。

SPACE PLANNING | 空间规划

多种多样的拱门和曲面为这个现代城堡构成了极具吸引力的流线型外观结构。一共有两层，其空间规划的特点在于门厅、客厅、餐厅等空间都有两层挑高，因此显得极为奢华开阔。尽管这座别墅很大但设计师并不打算填满它，而是选择简洁、明亮和轻快作为空间的代名词。

主卧：

　　舒适的床置于空间一端，床品、地毯等均选用淡雅的花色和精致的布艺，彰显出业主高贵典雅的情操。天花板是定制的手工雕刻木质造型，饰以菱形图案，使房间更加华丽大气。此外还有一个下沉区域，组合沙发、单椅、脚踏等应有尽有，大面积的落地窗，增加了采光度和通风性，在这里，或休息、或阅读、或品茶、或赏景，岂不怡然自得。

次卫：

　　蓝色的马赛克吊顶、浴池等，在整个住宅中展示了一股独特的小清新风。试想小巧的浴池中感受瀑布洒下的潺潺流水，一天的疲惫慢慢退去，身心得到完全放松。

主卫：

看上去像一个豪华舒适的公寓，拱形的天花板营造了一种开放宽敞感，尽显欧式的奢华与典雅。柱头、石膏线等雕刻精细，水晶烛台吊灯造型独特，时尚华丽。

家庭影院：

　　这个充满文化内涵的空间配备有高质量的音响和现代化的技术，可以让屋主、家人、客人享受到专业的电影院级别的质感。整个空间以神秘的黑色为主色调，搭配贵气的金色，弥漫着魔幻俏皮主义的色彩，为观影造势。金边黑色菱形地毯，不仅形成了视觉上的美感，也可以达到消音的效果，与黑色天花上零星的金点交相辉映。

THE EXOTIC STYLE
异域风情

DESIGN CONCEPT | 设计理念

The Casbah Bay is located in the Palm Desert in California, near the famous Bighorn Golf Club. This southern California Palm Desert home is inspired by the openness of Moroccan architecture and the carefree life. The palm trees in the yard make the interior and exterior a beautiful scenery. The rich furnishings, exquisite and complicated carvings and luxurious soft decorations make the villa gorgeous and comfortable, which is full of exotic feelings. In the boundless desert, having such a beautiful holiday resort must be very amazing. The owner herself is a Chicago-based real estate investor and a design aficionado. She wants to create a traditional Moroccan residence so that she visits a local architectural designer. Finally they complete the construction and design of this wonderful residence.

设计公司/设计师/摄影师/地点/面积

Design company : Mox Architecture
Location : Palm Desert, California, USA

Designers : Gordon Stein, The Owner
Area : over 1,765 m²

Photographer : Gibeon Photography

卡斯巴湾坐落于加利福尼亚州的棕榈沙漠，在久负盛名的大角羊高尔夫俱乐部旁。而这个南加州棕榈沙漠之家的设计灵感正是来自于摩洛哥建筑的开放性和无忧无虑的生活，庭院中的棕榈树使室内外都是一道靓丽的风景线。丰富的配饰，精致而复杂的雕花，豪华的软装，使整个别墅华丽而舒适，充满异国情调。在漫漫沙漠中有一处如此精美的度假区，想必是十分的美好。业主本身是芝加哥不动产投资人和设计爱好者，她想要在沙漠中建造一个传统的摩洛哥住宅，于是，她拜访了当地的建筑设计师，最终完成了这个让人眼前一亮的住宅的建造和设计。

厨房：

　　厨房为开放式的设计，并遵循了室外起居空间的曲线，双开门的设计为进出空间提供了便利。业主用精雕的红木定制橱柜隐藏了冰箱和冷冻机，既节省了空间，又达到了美观的效果。此外，业主还设计了一个造型优美的中央岛区。地板铺陈着黑白格子地砖，营造了一种灵动的美感。

餐厅：

　　餐厅设有钢化玻璃落地门，朝向厨房的室外平台，增加了空间的采光。房间内深浅色调的对比十分引人注目，黑与白之间创造了一种视觉美感。棉麻帐篷设计的天花板，十分独特，悬挂着的当代镂空铁艺吊灯，时尚大气。

家庭活动室：

　　这个空间弥漫着浓浓的摩洛哥风格的特色，墙面被黑肥皂擦洗后看起来就像是大理石花纹皮革，给人一种魔幻俏皮的色彩感。壁炉的表面使用了玉色的水石灰石膏，时尚雅致。床榻前是一个小水池，可供孩子嬉戏玩闹。家庭活动室通往拱廊，那里有拱门、耶路撒冷大理石地板以及彩色护墙板，增强了摩洛哥风格的美感。

主卧：

　　主卧选用香槟色为主色调，营造了一种淡雅、舒适的氛围。设计师在卧室墙面使用了古老的表层处理技巧，给人一种复古的精致感。床头壁龛运用了传统的手法，精雕细琢、形状各异、错综复杂、环环相扣的图形，为灰泥板墙面打造了一个带花边的表层，整个墙面既时尚又秀美。

主卫：

以白色为主色调，加上不同几何图形的点缀，使整个空间线条清晰、干净利落。主卫朝向平铺的室外淋浴和私人花园，设计师专门设计了一个木漆屏风，以便让光线射入到这个相对私密的卫浴空间，浴缸也是定制的。

SPACE PLANNING　｜　空间规划

　　这个住宅折射着一种摩洛哥利雅得式风格，大部分的房间、书房或办公室都面向围绕着大型中央庭院的拱廊，整个建筑的边缘呈弧形，大理石柱均匀的矗立着，户外还有一个大游泳池，周边有很多棕榈树，景色非常怡人。

INHERITING FRENCH CLASSIC
传承法式经典

DESIGN CONCEPT | 设计理念

The design of the villa is French style. The exterior façade of the architecture adopts arches and Rome pillars which are made of huge-scale stones such as limestone. The main color of the façade is creamy white, which is magnificent and majestic. For the interior furnishings, the owner requests a traditional home with classic details and materials. The interior ceilings, the solid limestone columns, custom iron works, archaistic art droplights, wall paintings and sculptures from the Renaissance period and the exquisitely carved plaster lines convey the elegance of classical French style. At the same time, the designer adds sofas and dining tables and chairs with clear lines into the interiors, which reduces complicity of classical style and increases simplicity and freshness of modern style.

设计公司/建筑设计/室内装饰/摄影师/地点/面积

Design company : Trapolin-Peer Architects
Photographers : Photographs © Alan Karchmer/Sandra Benedum
Architecture : Peter M Trapolin
Location : New Orleans, Louisiana, USA
Interior Decorating : Alexa Hampton
Area : 1,115 m²

　　别墅设计属于法式风格，建筑外立面采用拱门和罗马柱造型，以石灰岩等大体量的石材建设而成，外观以米白色调为主，大气庄严。在室内装饰上，屋主要求设计师为他提供一个由经典细节与材料打造的传统之家，室内穹顶天花、立体石灰岩大理石柱与定制的铁艺品，仿古艺术吊灯，文艺复兴时期的壁画和雕塑，精美雕刻的石膏线，传递出古典法式的优美。同时设计师在室内加入线条分明的沙发和餐桌椅，少了古典风格的繁重，多了一份现代风格的简洁和清新。

前门廊：

  前门廊保留了经典设计传统，比如石灰岩三角楣饰和科林斯石灰岩柱，其支架和装饰冠石均按照建筑师提供的草图手工雕刻，另外所有铁制品都是定制设计的。门廊运用了结实的多利安石灰岩和爱奥尼亚圆柱，并在其中使用了类似铁轨的图案。

门厅：

　　入口门厅位于东西轴线上，连接了书房和餐厅，这也是南北轴线的起点。圆形大厅是房子的核心，它连接了主要的公共空间，同时形成了一二层阶之间的关联。坚固的手工雕刻石制纵梁和石制金属保护层由悬臂式钢管承托，大型的壁柱和壁龛不仅是装饰元素，在房屋的力学系统中也扮演了尤为重要的角色。

SPACE PLANNING | 空间规划

别墅分为三层，进入房屋首先会注意到从正入口门厅到后露台及后喷泉壁之间的南北轴线。入口门厅沿东西轴线排开，贯穿书房和餐厅，让人们观赏到水池和喷泉壁的景色。房子背部的一条东西轴线，从带顶过道到游泳池，通向花园、后廊和客厅。从圆形大厅到教堂花园，从厨房到壁泉，更多的轴线景观，几乎都可让屋主体会到与外部环境的紧密联结。别墅的设计将所有公共生活空间设置在一楼，卧室则在二楼。空间都是沿着房子前后轴线或长度轴线设立，房子中央是一个多层圆形大厅，将一楼的公共空间与二楼的私人空间连接在一起。一些附属建筑及水池经由一条带顶的过道与房屋主体相连，扩展了后方的轴线，其他轴线通过景观元素的布置在视觉上得以延伸或被用于展现重要的艺术品。

主卧主卫：

　　主卧套房占用了二楼一半的面积，传统设计与经典细节相结合的同时还运用了现代便利设施。一部液晶电视隐藏在床对面的单面镜之后，屋主可以通过靠窗座位处的DVD播放器观看电影。主卫虽然很开放，但拥有独立的男女分区，该套房还设有一间小咖啡屋、独立的起居室，以及男女衣柜。

ELEGANT QUALITY LIFE
优雅品质生活

DESIGN CONCEPT | 设计理念

The villa is situated in Montecito, California. The designer shows his talent by designing this villa, and he is inspired by the designs of Andrea Palladio who is an Italian master architect in the period of Renaissance. Designed in a neo-classical style, the villa has an impeccable architectural proportion and structure, which is full of beauty, sunken in the surrounding landscape and punctuated by the infinity-edge swimming pool which precedes the polo field. Belonging to one of the prominent international hoteliers, the owner wants a mansion with horse stables and a polo field. The design is integrated into the existing land topography to accommodate the spectatorship of the polo game. The Villa is designed in such a way that the guests can enjoy watching the polo game match, atop the patio with the infinity edge pool, or on the terrace near the fireplace.

设计公司/设计师/摄影师/地点/面积

Design company : Kazakov Design
Location : Montecito, California, USA

Designer : Dmitriy Kazakov
Area : 1,858 m²

Photographer : Isa Totah

这套高级别墅位于加利福尼亚州的蒙特西托，设计师通过设计这套别墅，展示了自己非凡的设计才华，设计灵感则来自于文艺复兴时期意大利建筑大师安德里亚·帕拉第奥的设计。别墅的设计定位为新古典主义风格，有着完美无缺的建筑比例和结构，充满美感，周围有美丽的自然风景，最大的亮点是马球场前面的无边游泳池。屋主是著名的国际酒店经营者，他希望拥有一座带有马厩和马球场的大住宅。因此，为了方便观看马球比赛，设计师在设计时充分利用和结合了现有的地形地貌。别墅以这种方式设计，可以让客人在带有无边游泳池的露台或壁炉旁的平台上，享受观看马球比赛的乐趣。

客厅：

整个双客厅空间以大面积的落体窗帘为主体，设置于一角的钢琴，营造了充满人性与文艺的亲切氛围。独特的水晶吊灯与线条感十足的现代天花形成呼应，展示了设计师精湛的雕刻技术。方形的印花地毯、单椅、组合沙发等，纹理丰富，质感舒适，为屋主提供了良好的会客环境。

入口大厅：

　　以穹顶凹陷式的入口来迎接客人，既可以欣赏到该地产的全景，也可以到达室内的各个区域，动线之流畅由此可见。前厅由靠墙大理石柱均匀支撑，雕刻精细，对称分布的壁画惟妙惟肖，搭配椭圆的吊顶，步入前厅，颇有一种进入殿堂的感觉。

餐厅：

以皮质、流线型的餐桌椅为主，桌面有序地摆放着餐具，时尚简洁。红色的餐椅营造了一种温暖的氛围，使用餐成为家人之间共同度过的美好时光。在这个偏小的空间里，灯饰选用了晕黄的水晶吊灯，天花上的花枝图案与落地玻璃窗上的半透明图案交相辉映，营造出一种朦胧的美感。

SPACE PLANNING　|　空间规划

　　整个别墅占地面积大，各个空间的规划设计十分详尽，可见屋主对高品质生活的追求。主层按功能性分为活动区、休息区和工作区，户外还有游泳池、车库、汽车旅馆等。步入较低一层，会看到一个壮观的私人娱乐室，各种娱乐休闲设备应有尽有。从较低一层出来，在宽广的英式花园中觅得一抹祥和，这里有来自世界各地的植被、多样化的池塘、私人锦鲤池，还有复古的大理石雕像。

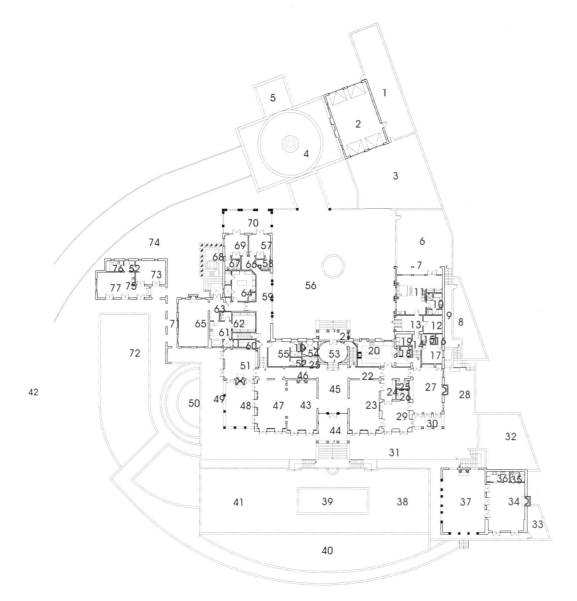

FLOOR PLAN 平面图

1. COMPRESSOR GENERATORS AREA
2. LIMO GARAGE
3. LIMO MOTOR COURT
4. OUTER MOTOR COURT
5. WESTERN ACCESS DRIVE
6. GUEST GARDEN
7. NORTHWEST PORCH
8. MECHANICAL/ELECTRICAL AREA
9. NORTH WALKWAY
10. GUEST BATH
11. GUEST HOUSE
12. EQUIPMENT
13. SERVICE PORCH
14. SERVICE HALLWAY
15. MAID'S BATH
16. MAID'S CLOSET
17. MAID'S ROOM
18. PANTRY
19. LAUNDRY
20. KITCHEN
21. PORCH WEST
22. HALLWAY
23. DINNING ROOM
24. BUTLER PANTRY
25. POWDER
26. BAR
27. FAMILY ROOM
28. BBQ TERRACE
29. SOLARIUM
30. NORTHEAST PORCH
31. NORTHEAST TERRACE
32. UPPER BBQ TERRACE
33. POOL EQUIPMENT
34. GUEST POOL HOUSE
35. GUEST POOL BATH
36. GUEST POOL KITCHEN
37. POOL LOGGIA
38. NORTHEAST POOL TERRACE
39. EAST POOL TERRACE
40. POOL
41. SOUTHEAST POOL TERRACE
42. SOUTHEAST TERRACE
43. LIVING ROOM
44. EAST PORCH
45. ENTRY HALL
46. HALLWAY
47. SITTING ROOM
48. SOUTH PORCH
49. SOUTH TERRACE
50. SOUTH POND
51. LIBRARY
52. CLOSET
53. ENTRY VESTIBULE
54. HALL
55. GUEST BEDROOM
56. INNER MOTOR COURT
57. ASHTON BEDROOM
58. ASHTON BATH
59. GALLERY HALL
60. FAX ROOM
61. HER BATH
62. HER CLOSET
63. MASTER BEDROOM HALL
64. HIS BATH
65. MASTER BEDROOM
66. CHILDREN'S HALL
67. MADISON BATHROOM
68. KOI POND
69. MADISON BEDROOM
70. CHILDREN'S COVERED PORCH
71. UPPER/LOWER LEVEL SOUTH WALKWAY
72. SOUTH POOL TERRACE
73. BEDROOM
74. GUEST PARKING
75. ROSE GARDEN GUEST HOUSE
76. BATH
77. LIVING ROOM

1、压缩发电机房
2、豪华车库
3、豪华汽车旅馆
4、外部车场
5、西侧行驶入口
6、客房花园
7、西北门廊
8、机械电器房
9、北通道
10、客卫
11、客房
12、设备间
13、机房门廊
14、机房玄关
15、保姆卫生间
16、保姆衣帽间
17、保姆房
18、餐具室
19、洗衣房
20、厨房
21、西门廊
22、走廊
23、餐厅
24、管家餐具室
25、女卫
26、吧台
27、家庭活动室
28、自助烧烤阳台
29、日光浴室
30、东北门廊
31、东北阳台
32、上层自助烧烤阳台
33、泳池设备间
34、客房泳池
35、客房泳池卫浴
36、客房泳池厨房
37、泳池凉廊
38、泳池东北阳台
39、泳池东阳台
40、泳池
41、泳池东南阳台
42、东南阳台
43、客厅
44、东门廊
45、入口大厅
46、过廊
47、起居室
48、南门廊
49、南阳台
50、南池塘
51、书房
52、衣帽间
53、入口前厅
54、过厅
55、客卧
56、内部车场
57、阿什顿卧室
58、阿什顿卫浴
59、长廊大厅
60、传真室
61、女卫
62、女衣帽间
63、主卧室大厅
64、男卫
65、主卧室
66、儿童房大厅
67、麦迪逊卫浴
68、锦鲤池
69、麦迪逊卧室
70、儿童房带顶门廊
71、南上下步梯
72、泳池南阳台
73、客卧
74、客房停车场
75、客房玫瑰园
76、客卫
77、客厅

137

家庭办公室:

红棕色实木打造的空间给人一种浓厚的文化气息,搭配深红的印花地毯,为屋主创造了一个良好的家庭工作环境。办公桌设计了隐藏的抽屉,既节省了空间,也起到了收纳的作用。实木橱柜里摆放着屋主钟爱的各种物品,或是自己收藏的,或是朋友赠送的,或是获奖得来的,承载着屋主不同时期的心路历程。

MOUNTAIN VILLA
优山大墅

DESIGN CONCEPT | 设计理念

Villa Ascosa (Hidden Villa) is aptly named, for it is truly hidden on a magical and spacious hill top setting. The villa is nestled at the top of one of the highest hills in Austin, offering endless uninterrupted views of the gorgeous Hill Country. The Italian estate consists of 4.5 acres at the end of a private gated street. An assortment of fruit trees and Italian cypresses have been planted on the property to add that Italian flavor. The villa is built to honor the traditions of old world construction using new world technologies. The 12 inches thick walls of the villa are built from aerated concrete blocks for maximum energy efficiency. The architectural design is created and chosen to nestle into the land with the look of passing time. No detail has been left unexamined for authenticity, function and beauty.

设计师/摄影师/地点/面积
Designer : Renata Marsilli
Location : Austin, Texas, USA
Photographers : Paul Finkel, Coles Hairson
Area : 929 m^2

阿斯科萨别墅（又名隐山别墅）名副其实，的确隐藏于一个神奇而宽阔的山峰处。别墅依偎于奥斯丁一座最高山的峰顶，一览美丽山区的景致。这个意大利式庄园位于一条私人的封闭街道，有1.82公顷之广，栽有各式各样的果树和意大利柏树以增添意式风味。别墅以纪念旧时代的传统建筑物为原则，采用新时代技术修筑。0.3米厚的墙用加气混凝土块切成，使能源效率最大化。所选用和创建的这个建筑设计见证着时间的流逝。所有细节都经过检验，只为实现其真实性、功能性和美感。

客厅：

　　繁简有度大概是设计师奉献给屋主最好的礼物。简约舒适的沙发，精致繁复的花纹地毯、抱枕；透亮清爽的水晶，古朴优雅的铁艺；少许绿植、花卉的置入更显生活气质。

　　手工混合的岩石和灰泥被用于外墙饰面。马赛克大理石、精致的平线脚和耶路撒冷石地板等一系列材料的混搭使用，令室内的设计变得更加优雅。石头、青铜壁挂喷泉和贝壳壁灯互相呼应，营造出宜人的空间氛围。独一无二的吊灯给奶油色石膏墙镀上一层温暖的光晕。

餐厅：

纯正的欧式风让简约的白、灰色调作为家居的主调，简单的颜色，豪华舒适的效果，再加上适当点缀的金色线条，空间感强烈且熠熠生辉。窗外大片的蓝透过大面集落地窗辐射入室内，豪华感的水晶吊灯与纯净的鲜花，整体空间奢华而浪漫，屋内仿佛吹着一股典雅浪漫的欧洲风。

LOWER FLOOR PLAN 底层平面图

1. POOL	11. MECH ROOM	1、泳池	11、设备室
2. SPA	12. STUDIO	2、温泉浴场	12、画室
3. POOL TERRACE	13. MECH/AV CLOSET	3、泳池露台	13、机械/音响柜
4. STUDY	14. WINE CELLAR	4、书房	14、酒窖
5. LOGGIA	15. GALLERY	5、凉廊	15、走廊
6. VESTIBULE	16. MEDIA	6、门廊	16、媒体室
7. BEDROOM	17. OUTDOOR LIVING	7、卧室	17、室外客厅
8. BATH	18. LOWER TERRACE	8、浴室	18、楼下露台
9. EXERCISE	19. PERGOLA ARBOR	9、健身房	19、绿廊藤架
10. CLOSET	20. TERRACE	10、更衣室	20、阳台

厨房：

独一无二的灰泥墙壁打造出一个干净透明的厨房空间，开阔的格局动线让中岛吧台也显得宽大起来。简单精细的稍加修饰，让雕花、绿植成为空间的主角，仿佛生命在跳动，味蕾在舒张！

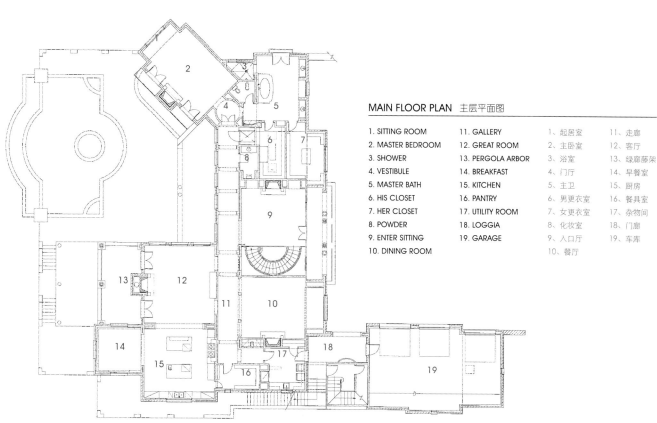

MAIN FLOOR PLAN 主层平面图

1. SITTING ROOM	11. GALLERY	1、起居室	11、走廊
2. MASTER BEDROOM	12. GREAT ROOM	2、主卧室	12、客厅
3. SHOWER	13. PERGOLA ARBOR	3、浴室	13、绿廊藤架
4. VESTIBULE	14. BREAKFAST	4、门厅	14、早餐室
5. MASTER BATH	15. KITCHEN	5、主卫	15、厨房
6. HIS CLOSET	16. PANTRY	6、男更衣室	16、餐具室
7. HER CLOSET	17. UTILITY ROOM	7、女更衣室	17、杂物间
8. POWDER	18. LOGGIA	8、化妆室	18、门廊
9. ENTER SITTING	19. GARAGE	9、入口厅	19、车库
10. DINING ROOM		10、餐厅	

廊道：

细腻精致是行走在廊道上的第一感受。拱形的圆弧交叉设计，配以精细的花纹，古铜吊灯与壁灯填补了空间的历史印记，造型弯曲有度，天使相护，就如漫步在时间的回廊里。山水相依，相逢有时！

SPACE PLANNING | 空间规划

这是一个隐藏于奥斯汀山古老乡村的住宅，所处的地理位置非常隐蔽。西湖区离奥斯汀市中心最短需要10分钟的路程，并且邻近购物和娱乐中心，使其成为理想的住所。屋主希望空间在满足物质生活的同时更注重心灵的慰藉，因此一切规划以此出发，一应俱全。值得特别称赞的当属户外泳池和书房，身体和心灵总有一个在路上。

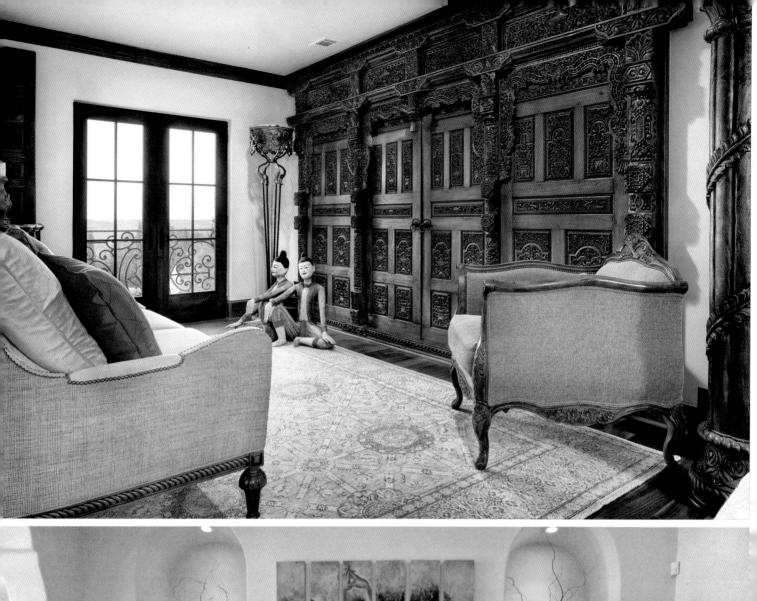

卧室：

　　扩大别墅天然的地理位置优势，四方格子玻璃窗将窗外的风景切成一块块拼图画，有趣而具文艺气息。简约的床品和家具，更加符合气质的需求。淡淡的蓝调，给房间披上一层梦幻的色彩。

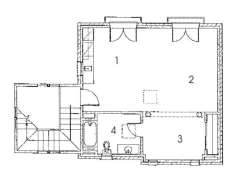

CASITA FLOOR PLAN　阁楼平面图

- 1. KITCHEN/DINING　1、厨房/餐厅
- 2. LIVING ROOM　2、客厅
- 3. BED ALCOVE　3、卧室
- 4. BATH　4、浴室

酒窖：

缤纷的色彩由天然的灰石组合而成，优美的弧线让空间更显立体，纯黑的铁艺吊灯与壁灯突出了空间的厚重感，长长的沙发椅舒适又亲密，而古老的头像壁画让酒窖文化更具质感。

A HILLTOP RESIDENCE WITH PERFECT VIEWS
绝佳观景，山顶美宅

DESIGN CONCEPT | 设计理念

Situated on a hilltop, the residence has commanding views, east from Santa Monica to Point Dume in the west, as well as north to the Santa Monica Mountains. The design challenge here is how to approach the residence from the view side facing toward the Pacific Ocean and not have that be the front door. After discussions with the owners and morphological analyses, the designer comes up with a new thought by creating the arrival between two one story garages that leading to the entry garden and pergola to achieve the goal of further enjoying the beautiful scenery in Pacific Ocean. This is a design that exemplifies individuality, created over time rather than in a single, sweeping stroke. Bringing together the tastes and eccentricities of the owners, the designer makes the Malibu house a warm invitation into their life as if they are bathed in the bright Californian sunshine, which is free and cozy.

设计公司/设计师/摄影师/地点/面积

Design company: Paul Brant Williger, Architect
Location: Malibu, California, USA
Designer: Paul Brant Williger
Area: 1,300 m²
Photographer: Nick Johnson

本案坐落在山顶上，呈居高临下之感，东至圣塔莫尼卡、西至庞特度姆、北达圣塔莫尼卡山。设计师面临的挑战是如何不运用前门便可使本案临近太平洋景观。经过与主人的连番讨论和分析地貌，设计师另辟蹊径，通过打造两个分别朝向入口花园和绿廊的车库，得以实现进一步观赏太平洋美景的意愿。这是一个彰显个性的设计，是经过一段时间创造出来的，而不是一次单一的以偏概全的尝试。设计师将屋主的品位与喜好充分结合，使马里布之家成为了他们温馨生活的载体，如沐浴在加州温暖的阳光中，自在慵懒。

正式客厅：

外露的木梁、石材地板以及大量木制品奠定了客厅的整体基调。但客厅整体氛围也不是一成不变的，开放型的空间设计让外界的景观能成为整体空间的一部分，把自然景观、海景、蓝天等都纳入室内环境范围。随着时间、季节的变化而不断变幻着不同的景色。再加上人的参与，使空间极富动感，真正体现了空间"共享"的特性。

待客区：

　　待客区圆顶天花不同于常规天花板设计，慵懒的幅度透露出随和的基调，曲面的天花板形式更富于变化，整个空间是流动、畅通的。与优美的多利安拱门搭配在一起，形成美的空间构造，为待客区增添了欢快的韵律。白色调的选择，让整个空间干净、轻松。大幅落地窗的窗架与玻璃茶几的边线相呼应，简洁的现代化感觉浓厚。纤细线条的吊灯精致婉约，与整体氛围更为搭配。米黄色沙发的选择则为空间增添了静谧、柔和气质。

SPACE PLANNING ｜ 空间规划

　　马里布之家所在的山顶地形，拥有绝佳的自然风光和观景角度。为了让屋主能不受一般房屋结构限制，自在欣赏景观，设计师打破传统格局的局限，用开放式的设计手法、简练的线条、淡雅的色彩，打造一个舒适自在的观景空间。无论是在正式客厅，还是待客区，抑或是卧室、餐厅，都不会错过室外的美景。同时马里布之家各个宽绰的空间之间无缝衔接，屋主可在每个空间切换自如，如同在原野般自由而通畅，为居住者提供了戏剧化的舞台。

卧室：

　　每种看似不起眼材料和摆件的运用，都有可能影响到空间的整体质感。材料的搭配，是凸显空间氛围的重要方式。同时，家具、摆件的选择也可看出屋主的喜好和品味。多种多样的小狗摆件说明了屋主对小狗的喜爱，希望它能时时陪伴在屋主身边。房屋其他空间也可见到小狗雕塑，家里也饲养了小狗。小狗不再是简单的宠物，而是成为了家人朋友般的存在，能够与人互信陪伴，是感情的寄托。树干造型的储物架也值得称赞，纹理逼真。为了与树干造型搭配，台面做成粗糙的木纹和仿木质的颜色，表现了屋主对原生态的追求和向往。桌面上的绿色盆栽，特意修剪的圆形造型质朴可爱，有让人不自觉寻找生命和自然的亲近之感。这里是屋主的避风港和最坚固的港湾，在这里，可以放下所有，静静地享受这一方小天地带来的静谧。

COLOR MATCHING ｜ 色彩搭配

棕色、金色和米黄色一同构成了空间的主色调，三种颜色组成的柔和色调完美框定住了户外的自然景色。使得屋主既能感受到轻松温和的室内氛围，也不会错过活泼的室外景观。深与浅，浓与淡，明与暗，形成对比，使得马里布之家没有一处是沉闷无趣的，丰富的色彩变化带给屋主多重的居住体验。

SOFT DECORATION ｜ 软装搭配

装饰方面，设计师则选用了传统与现代家具组合的方式，通过不同家具方位、排列的精心搭配，形成不同的空间气质。利落的直线条橱柜、茶几与柔和的弧形沙发相搭配，既有现代气息的干净利落，也不乏传统风格的典雅大方。这种风格和材质上的折中混合，穿越了时空，为这栋别墅增添了慵懒的氛围，给那些非常正式的空间带来地中海的柔情。

OLD WORLD CHARM
古雅世界的魅惑

DESIGN CONCEPT | 设计理念

The European inspired home is designed to create a cozy and comfortable living atmosphere for the owner. The designers pay much attention to the structure itself and make the owners re-experience the coziness and comfort when traveling abroad by details. The original structure in European style reflects the owners' respect and love to nature. The design pursues a sense of spiritual belonging and gives people a fresh natural breath. So the designers keep the exterior features of the residence. The plaster walls collocate with the walls piled by archaized bricks, which has a unique charm. The distinct outdoor garden and swimming pool add fun to owners' outdoor life. The simple roofed outdoor dining room and BBQ area are good places for owners and friends to party and entertain. Based on these characteristics, the designers design and recreate the interior to add flavors and cultural temperament to this antique house.

设计公司/摄影师/地点/面积
Design company : IMI Design
Area : 575 m²

Photographer : Dino Tonn

Location : California, USA

这座欧式风格的住宅旨在为业主营造一种安逸、舒适的生活氛围。设计师非常注重建筑本身的结构，通过细节的处理让业主重新体验旅居国外时的惬意与舒畅。欧式风格的原始建筑体现了业主对自然的崇尚及热爱，其设计追求心灵的归属感，给人一种扑面而来的自然气息。因此，设计师保留了建筑的室外特色，灰泥墙面搭配裸露的仿古石砖砌成的墙面，别有一份韵味；独特的户外花园和游泳池，为屋主的户外生活提供了乐趣；简易的带顶户外餐厅和烧烤区，是屋主与亲朋好友聚会、休闲的佳所。在此基础上设计师对室内进行了设计和改造，为这座古色古香的住宅增添了几分韵味与文化气息。

SPACE PLANNING ｜ 空间规划

这个住宅的建筑面积很大，根据业主的需要划分成了很多不同的空间，并对不同的空间进行了相应的设计以满足业主各方面的需求。主层的空间分布动静结合，错落有致，大厅、门厅和过道构成的轴线将动静区连接起来，使整个空间的规划更合理，也方便业主的日常生活。

客厅：

　　客厅天花采用开放式橡梁，由弯曲的铁十字架支撑，搭配简单古朴的铁艺灯，挑高的中空，尽显宽敞舒适。客厅的最大特色在于它的半开放式设计，推拉式大门收缩之后眼界完全敞开，户外的游泳池和远方的风景尽收眼底，这种半开放式的设计极大地增强了室内的采光和通风效果。

入口大厅：

　　几根大理石柱对称分布矗立，支撑着中空的圆顶天花，营造宏伟、庄重之美，天花上模仿意大利壁画的手绘形态悠然，十分灵动。地板以天花下方的供桌为圆心错落有致地铺设开来，四层弧形台阶的立面选用了不同的马赛克瓷砖铺陈，细节之处体现了设计师的匠心精神。

FLOOR PLAN - MAIN LEVEL
主层平面图

1. MASTER BEDROOM	1、主卧
2. MASTER BATH	2、主浴
3. HALL	3、大厅
4. MASTER CLOSET	4、主衣帽间
5. EXERCISE ROOM	5、健身房
6. LAUNDRY	6、洗衣间
7. HALL	7、大厅
8. POWDER ROOM	8、化妆间
9. STUDY	9、书房
10. BATH	10、浴室
11. GREAT ROOM	11、客厅
12. FOYER	12、门厅
13. FAMILY ROOM	13、家庭活动室
14. BREAKFAST ROOM	14、早餐室
15. KITCHEN	15、厨房
16. PANTRY	16、餐具室
17. HALL	17、过道
18. DINING ROOM	18、餐厅
19. WINE ROOM	19、藏酒室
20. BEDROOM	20、卧室
21. BALCONY	21、阳台
22. BATHROOM	22、浴室
23. CLOSET	23、衣帽间
24. STAIRWELL	24、楼梯间
25. LIVING ROOM	25、客厅
26. BALCONY	26、阳台
27. BEDROOM	27、卧室
28. STAIRWELL	28、楼梯间
29. CLOSET	29、衣帽间
30. BATHROOM	30、浴室

厨房：

　　厨房中的设计反映了不同时期的装饰风格和地方风情，中岛台上，抽屉把手为藤蔓雕刻，而亚之竹被精雕成终枝环绕四边，四角顶点以亚之竹心作为聚点，疏密对比收放得当，华丽的手工雕刻和手绘件与一些稍为简单的配饰相结合，达到了美学平衡。

书房：

　　私人定制的书桌独具匠心，四角立面雕刻着狮头和狮脚，是曲与直的刚柔并济；正立面结合橄榄绿皮革硬包，减少了实木满屋的厚重感，吊顶采用了与书桌相呼应的定制手法；细节处胜于在对角线的材质上选用了铆钉打造，复古的奢华感更加强烈。书房的色彩并不单调，木色与橄榄绿的搭配，提升了整个空间的质感，营造了沉思静悟、安顿心灵的阅读氛围。

CHARMING IMPERIAL HOME
迷情帝国之家

DESIGN CONCEPT | 设计理念

The overall interior design concept of the villa is one of elements conveying local regional influences and a reflection of the primary exterior design characteristics of this 19th century Russian imperial house. Illuminated by numerous wide windows, the entrance hall has floors combining Belgium black and Giallo Siena marble, and walls covered with gold & white stripe wallpaper. All of the wooden surfaces of the furniture are in white finish and gold leaf. The lighting elements have been designed and manufactured in line with the period of the building. The halls connecting various spaces incorporate the same materials and color palette as the entrance hall. The gold wall and sofa make the living room a resplendent and magnificent palace. The custom-made carpet, the interior furnishings and curtains collocate harmoniously. The classical crystal droplights, the ceiling lamps, the wall lamps and the owner's favorite classical paintings from the 15th century are the owner's sincere tribute to the classical humanistic feelings.

设计公司/设计师/摄影师/地点/面积

Design company : Eren Yorulmazer Interior & Architecture
Location : Batumi, Georgia, USA
Designer : Eren Yorulmazer
Area : 1,500 m²
Photographer : Ali Bekman

别墅室内设计理念是传达当地地区性风格的元素之一，也是19世纪俄罗斯帝国住宅原始室内设计特点的一种反映。大型窗户让门厅光线充足，门厅内的地板融合了比利时黑大理石和锡耶纳麻大理石等，墙壁覆有金色和白色条纹墙纸，而所有家具的木饰面则运用了白漆和金箔，此外，灯光的设计和制作符合了该建筑的时代特征。连接不同空间的大厅采用了相同的材料和色彩，体现了室内空间的整体性。金色的墙面和沙发让整个客厅显示出金碧辉煌的宫殿感，定制的地毯、室内装饰物品和窗帘的搭配协调一致，古典的水晶吊灯、吸顶灯和壁灯，屋主喜爱的15世纪古典壁画，都是屋主向古典人文情怀的衷心致敬。

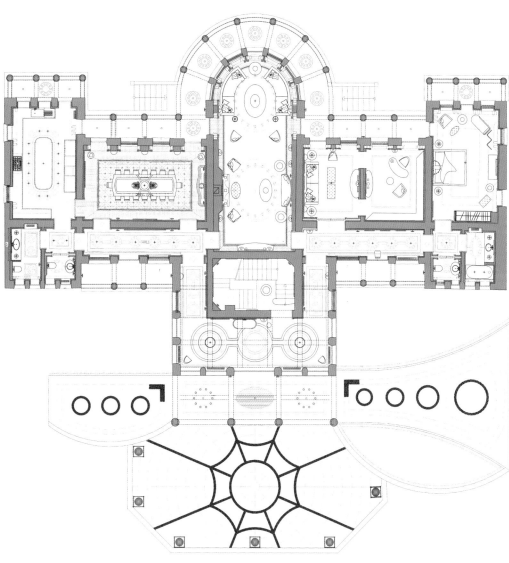

客厅：

整个空间的完整度取决于黄绿色调的鲁贝利窗帘和室内装潢以及定制的地毯和天花板装饰物，青铜灯具和木材表面反映了该建筑的时代特征。空间围绕椭圆咖啡桌成形，并通向起居室和壁炉区，白色壁炉前的红宝石色丝绸软垫金箔椅与相同颜色的天鹅绒面脚榻相互呼应。

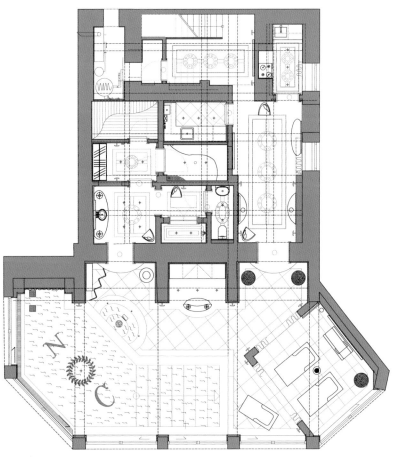

餐厅：

　　餐厅的中心是一张黑檀木桌，墙壁、窗帘和座椅都使用了猩红色布艺，一盏由流苏、水晶以及烛台组成的青铜吊灯让整个空间熠熠生辉。天花板装饰展现了时代特征，使用了通常用于墙面的黄色和蓝色，不仅凸显了细节特征还使墙壁与天花板互相融合。

SPACE PLANNING | 空间规划

　　楼梯间将上下两层的空间连通，使之协调，楼梯有木质扶手，墙面还有许多青铜饰物，阶梯上铺设的地毯不仅可以防止跌倒，还丰富了空间的色彩和纹理。一楼进门处偌大的门厅，门厅面向由白色、灰绿和黑色点缀的书房和图书室，客厅、家庭活动室、休闲室和餐厅厨房相连，展现了设计师流畅通透的空间布局理念。二楼以休息空间为主，一间宽大的主卧套房空间配套齐全，五间次卧则用不同的主题色彩展示其特色和个性，大面积的纯色搭配，让卧房呈现恢宏的气势和尊贵感。

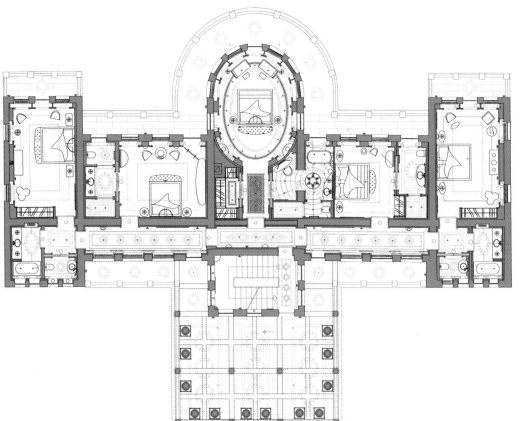

主卧：

主卧室的墙由淡褐色的丝绸棉锻覆盖，左右两侧分别是更衣室和浴室。室内的所有家具与墙面色调一致，床头板则以金箔饰边，窗帘与墙壁覆盖面采用了相同的材质，烛台和灯具为青铜制品，镶木地板则与建筑物特征保持了统一性。

客房：

　　客房延续了同层起居室的色调，黄色丝绸织物、墙面与红宝石色电视柜、木雕床、石油绿屏风互相搭配，其中还铺有色彩缤纷的地毯，水晶吊灯的加入让整个空间更加明亮。

客房：

　　另一客房则运用了紫色、绿色、黄色和白色，定制地毯和亮紫色的家具令空间更加富有生气。热情洋溢的配色延伸到窗帘和床罩上，成为视觉焦点。

　　红、黑、白搭配而成的客房折射出地域特征，展现了这一帝国传统宅邸的设计风格。

RENAISSANCE HOME
文艺复兴之家

DESIGN CONCEPT | 设计理念

The villa is designed by the famous interior designer Marie Peterson who has devoted her career to creating inviting spaces reminiscent of French chateaux, English country manors and Italian villas. Her painstaking attention to detail and flawless execution have earned her an exemplary reputation. The owners of the home are an enthusiastic couple who love Italian Renaissance style. They often entertain on a grand scale and utilize their beautiful property for a host of fundraising activities. The wife has a passion for Italian art and antiques and has acquired some truly important pieces during her travels. And the designer has put them in the house. The joy of this project is incorporating those wonderful objects d'art and furnishings throughout the home, decorating with rococo mirrors and textiles, and creating a rich palette to highlight them. The designer creates a space which is magnificent and majestic, and her professional vision and skillful control of scale make it a spectacular residence.

设计公司/设计师/摄影师/地点/面积

Design company : Chelsea Court Designs, Inc.
Designers : Marie Christine Peterson (Principal Designer), Sascha Marie Lale (Senior Designer & Project Management)
Photographer : David Livingston   **Location** : Monte Sereno, California, USA   **Area** : 929 m²

别墅设计出自著名室内设计师玛丽·彼得森,她擅长法国城堡、英式乡村庄园和意大利别墅的空间魅力展示,细致入微的设计理念和完美的执行力让她赢得了声誉。而本案屋主钟爱意大利文艺复兴时期的风格,屋主是一对热情的夫妇,他们经常盛宴款待宾客,并利用自己漂亮的府邸作为一系列筹集善款的地点。女主人热爱意大利艺术和古董,并在她的旅行中获得了一些重要的艺术品,于是放入家中。这个项目的乐趣是在整个家中贯穿式艺术品和家具,并采用洛可可式镜和大量的纺织品作为装饰物,同时运用丰富的色彩方案以凸显它们。设计师描绘的空间既华美又宏伟,她的专业眼光与对尺度的巧妙把握结合在一起,打造了一个蔚为壮观的居所。

客厅:

客厅以壁炉为中轴线,在庄严的空间中非对称的摆放着浓酒红的长条沙发和稍作雕饰的简约沙发椅,黄玉色花纹地毯和窗帘与墙面颜色相呼应。古老大理石壁炉上傲然陈列的是一面18世纪洛可可式风格的镜子,华丽繁复、雕饰线条纤细优美。

SPACE PLANNING │ 空间规划

别墅分为两层，以入门大厅的弧形楼梯为界，别墅被有序地规划为两层。一楼以公共活动区域为主，客厅宽大舒适，餐厅和下午茶的休闲室彼此相连，并设计了很多活动座椅，满足了屋主热情好客的需求。二楼以卧室和书房为主，主卧套房配有更衣室、化妆间和主卫，主卫占面大，两个盥洗台和一个靠窗浴缸的设计贴合屋主心意。二楼的设计和家具布置意在打造安静的休息空间，有助于提高屋主的睡眠质量。

餐厅：

　　餐厅以黑色和黄昏色为主调，延续了客厅亲密的氛围，墙体进行简约的雕饰，线条简明优雅，增加空间设计感又不至于太繁复。两个定制的方形桌，以黑漆喷涂、金边勾勒，进而提升聚会派对中的欢乐交谈氛围。宽大的法式落地玻璃门和柔软的真丝帘幔，成就了愉快的用餐氛围。

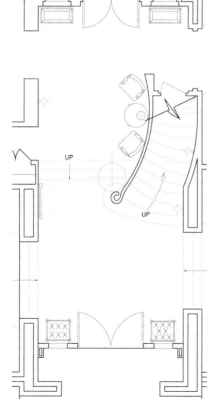
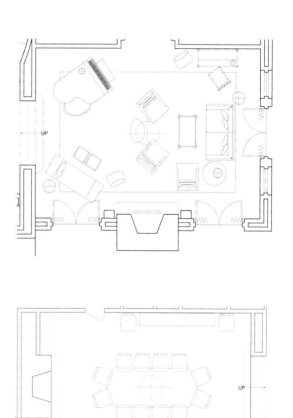

主卧：

华丽的金色贝加莫布料装饰着主卧室的墙壁，青色和嫩蓝色的布艺沙发椅和床品共同营造出一种柔和平静的氛围。茶黑色的四柱床被昂贵的丝绸帷幔包裹着，枯色帷幔和太妃奶糖色床品亲肤柔软，让卧室更显稳重华贵又具有浪漫气息。

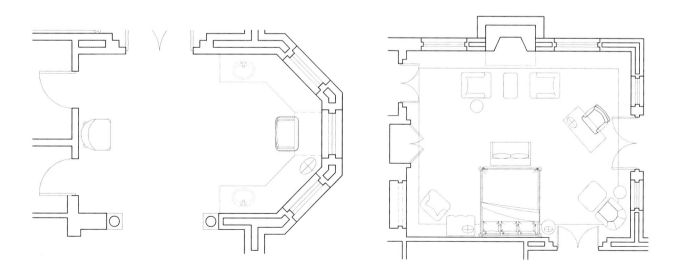

GOLD HOME
金色之家

DESIGN CONCEPT | 设计理念

The concept for this project comes from the old original houses that line the beautiful Bosphorus Sea. The main concept is to create daily living for all the family in a traditional and a contemporary setting. Harmonious volumes, carefully planned lighting, quality furniture and rich fabrics are the main points of the design. The designers use a lot of traditional materials to breathe a soul into the building. The combination of traditional ottoman motifs and Islamic prints creates a light and open space in a classical style and reflects the theme by interior decorations. The modernized ottoman motifs on the flooring, ceiling and mirrors are clearly evident in this house. On the premise of focusing on ornamental effects, the designers use modern decorative methods and new materials to restore the classical temperament, such as simple lines and modern materials in traditional designs. The designers pursue the general outline and features in classical style, and deduce the highest state of the neo-classical space by the way of "shape dispersing but spirit concentrating".

设计公司/设计师/摄影师/地点/面积

Design company : Sia & Moore Architecture and Interior Design
Photographer : Ersen Çörekç
Designers : Banu Altay, Kübra Altun, Esra Arga, Abdullah Yılmaz
Location : Istanbul, Turkey
Area : 650 m²

博斯普鲁斯海滨别墅项目的设计概念来自于旧宅两旁博斯普鲁斯海峡的美景，其主要的理念是为屋主全家人创造一个兼具传统与现代的日常生活模式。和谐的体积，精心策划的灯光，优质的家具和丰富的面料是设计的重点，同时设计师采用大量传统材料为建筑注入灵魂。传统的奥特曼图案与穆斯林图案相结合，以经典风格打造一个明亮开放的空间，并且用室内装饰来反映这个主题。地板、天花板及镜面上的现代化奥曼图案在这座房子里显而易见，在重视装饰效果的前提下，设计师运用现代装饰手法和新材质还原古典气质，简单的线条和现代材质的传统设计，在设计上追求古典风格的大致轮廓特点，以"形散神聚"的形式演绎出新古典空间的最高境界。

客厅：

白色软包沙发纯洁浪漫，黑色沙发经典雅致，沙发边角和茶几四面运用金色雕刻，彰显空间古典的贵族气息。天花以简约的石膏线做造型，中央的穆斯林绘画独具特色又和沙发椅上的雕饰相呼应。吊灯以水晶链条为装饰，昏黄的暖光与黑金色的窗帘在白色空间里演绎着新古典的韵味。

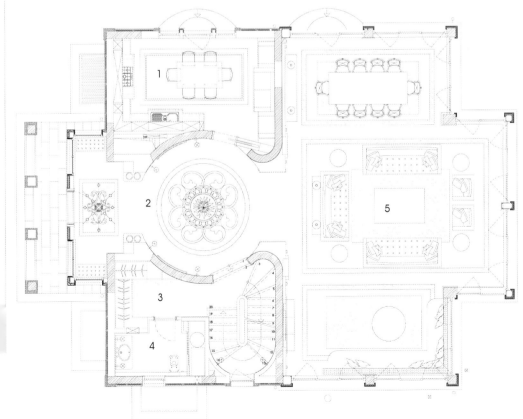

GROUND FLOOR PLAN 一层平面图

1. KITCHEN　　　1、厨房
2. ENTRANCE HALL　2、入口大厅
3. CLOAKROOM　　3、衣帽间
4. GUEST WC　　　4、客卫
5. LIVING ROOM　　5、客厅

餐厅：

为统一餐厅的走向和造型，餐桌、地毯和吊灯采用长方形设计，朝下设计的灯帽让光照射在通透的水晶上，温暖的黄昏色灯光让水晶吊饰熠熠生辉。三面大窗采光充足，黑金色的窗帘和台灯相呼应。餐桌椅、烛台和石膏线、餐边柜都采用黄金镶边，让别墅各空间都保持风格上的一致。

SPACE PLANNING | 空间规划

别墅分为三层，地下一层是一个休息区域，囊括了家庭影院、温泉以及健身房。圆形大厅富有设计感，主客厅、客厅、走廊和家庭影院的完美装饰展现着热情的氛围，华丽的木雕嵌板像艺术品一样陈列在主客厅和餐厅之内。一楼以公共活动区域为主，入门圆形大厅庄重，灵动的圆形设计让空间从三面分别连接厨房、客厅和弧形楼梯，餐厅和家庭影音室居于客厅两侧，成一条水平线设计，楼梯附近的卫生间和衣帽间是对空间的合理利用。二楼空间包括主卧套房和三间客卧，一楼双层入门大厅将空间拓展至二楼，因此二楼特意设计了一个圆形的走廊，在空间上与主卧圆形床呈对称分布。

FIRST FLOOR PLAN 二层平面图

1. BATHROOM — 1、客卫
2. BEDROOM — 2、卧室
3. GALLERY — 3、走廊
4. SHARED BATHROOM — 4、公共卫生间
5. STORAGE — 5、储藏室
6. HALL — 6、过厅
7. DRESSING ROOM — 7、更衣室
8. MASTER BEDROOM — 8、主卧室
9. ENSUITE BATHROOM — 9、套内卫生间
10. SHOWER — 10、淋浴房

主卧：

　　主卧室蕴藏着一切奢华元素：照明设备、金箔镶边的红色天鹅绒床头、精致的布艺床品，还有温和的奶油色与红色相融。卧室以古典精髓构造，同时融入现代与传统的碰撞，四根罗马柱显尽屋主尊贵的气势，被设计成为真正的私人领域。

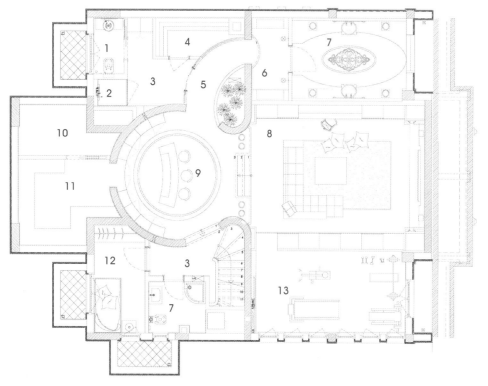

BASEMENT FLOOR PLAN 地下层平面图

1. SPA WC — 1、温泉卫生间
2. SHOWER — 2、淋浴房
3. HALL — 3、过厅
4. SAUNA — 4、桑拿房
5. SPA HALL — 5、温泉大厅
6. BATH HALL — 6、浴浴大厅
7. BATH — 7、卫生间
8. THEATER ROOM — 8、影音室
9. HALL — 9、大厅
10. WATER TANK-BOOSTER — 10、水箱增压器
11. PANTRY — 11、餐具室
12. SERVICE ROOM — 12、机房
13. FITNESS — 13、健身房

PECULIAR THOUGHT AND QUIET DREAM
绮思幽梦

DESIGN CONCEPT | 设计理念

Nestled on the pristine Fort Washington golf course in Fresno, California, this authentic Spanish villa is an architectural masterpiece. It reflects not only the "lifestyle" of the family, but also a "Spanish lifestyle" in the overall design process. Leisure, calm, funny, romantic and unrestrained life thoughts are penetrated into the Spanish style villa which is the best carrier of this kind of lifestyle. The diverse and distinct Spanish cultural connotation has an influence on all aspects of the design, such as classical flavors of Roman architecture and exotic elements of Oriental culture. The designers combine these different features together to create a villa with a unique temperament.

设计公司/设计师/摄影师/地点

Design company : Professional Design Consultants
Photographers : La BellaVita Photography, DJ Ellis, Travis C. Birchfield
Designers : Linda Zoerb ASID, CCIDC, Aneesah Sabree, DeAnn Martin
Location : Selma, California, USA

这座地道的西班牙风格别墅坐落在夫勒斯诺市一个古朴的华盛顿堡高尔夫球场之上,是一项建筑杰作。在整个设计过程中,它反映的不仅仅是这个家庭的"个性方式",也是"西班牙生活方式"的传达。闲适、从容、意趣、浪漫、奔放的生活意念,始终贯穿其中,这座西班牙风格别墅就是这一生活方式最好的载体。西班牙多元、独特的文化内涵也影响到设计的方方面面,既带有罗马建筑大方古典的韵味,又随处可见东方文化的异域元素。设计师将这些原本截然不同的特质糅合在一起,形成整座别墅独一无二的气质风情。

客厅:

圆形拱门是西班牙住宅的标志性符号,这座地道的西班牙别墅当然也不会忽略这个重要特点。马赛克装饰的圆拱门连接两个开阔流动的空间,会让观者产生一种奇妙的时空延伸感。高悬的圆形大吊灯一出场就夺人视线,让人不自觉驻足仔细研究。高低、大小不一的圆柱形灯筒为原本朴实无华的灯座增加了观赏性,线条圆润柔和的气质展露无遗,与烛光相互呼应,温暖、踏实之感充盈整个空间。细链条与粗重灯座形成对比,增强了趣味性。墙角的棱角被处理成圆形,独特的圆弧设计,让原本单一的墙体变得丰富起来,加上高大平缓的坡顶、不对称布局的墙面,犹如一位西班牙女郎和着弗拉门戈的节拍旋转飘起舞裙,透露出来的是柔和、开朗的建筑美感。加上本来厚重的墙体,给人安全、踏实的感觉,提高了居住的舒适度。

餐厅：

正式用餐区域内，散发着教堂般肃穆的氛围，这主要是受特别设计的圆顶天花板的影响。如夜幕般地暗色与皎洁无暇的白色线条共同勾勒出圆顶天花板的弧度，搭配造型简单的光面木质餐桌，颜色为统一的褐色，与瓷砖颜色相呼应，庄重、肃穆是这个空间的主基调。

SPACE PLANNING ｜ 空间规划

设计师以功能为标准将室内空间划区，充分体现主人优先的原则。二楼的私密区和一楼活动区动静分离，互不干扰。尊重个人的生活空间的私密性是尤为重要的，这表现在主卧套房不仅内置宽大的步入式衣柜和定制的法式衣橱，还备有独立的卫生间和浴室，保证私人空间充足而不受干扰，舒适而自得其乐。客厅采光性极佳，是与家人、朋友聊天聚会的场所，应在房屋的最好位置，体现了以居住者为本的功能性。

COLOR MATCHING | 色彩搭配

　　整体空间质朴、大方、自然，如客厅、餐厅多采用比较原生态的褐色和土黄色，融入了阳光和活力，体现了西班牙风情的质朴内涵和浓郁情怀。意大利金黄色的向日葵、法国绵延不断的紫色薰衣草花田、地中海碧蓝的天空等是这些西班牙符号化色彩的来源。这些浓烈而跳跃的色彩，巧妙地搭配在一起则显示出一种西班牙特有的情调美。作为休息区的卧室则有不同的选择：主卧套房选用米白色墙面搭配青色沙发、靠枕，呈现出淡然、安宁的氛围；儿童房其中一间的粉紫色带来温柔、沉静之感，另一间的格子花纹则素雅清新。

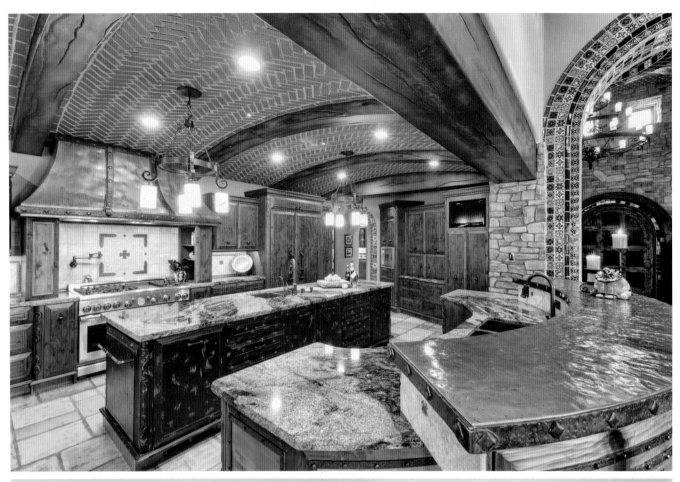

卫生间：

　　卫生间设计也是这座地道的西班牙别墅不可忽略的地方，它独特的西班牙风格特点尤为突出：绚丽、跳脱的颜色，异域的马赛克装饰，抽象化造型洗手台和马桶设计，装饰物随机的创意组合等。镂空的球体挂件闪耀出瑰丽的光芒、十字架造型的毛巾架宗教意味浓厚、绿色植物的摆件为空间增添了自然的清新感觉……每个细节都尽显西班牙风格的个性和魅力，向我们展现出西班牙民族热情、奔放和狂野的个性，充满了丰富的想象力和浪漫情怀。

AN ARTISTIC HOME
艺术家

DESIGN CONCEPT | 设计理念

Located at the corner of 79th Street and Park Avenue in New York City, the apartment is the home of members of one of the wealthiest families in New York. It is a family apartment where they have raised two children from his previous marriage and two from theirs. The couple is not only social, but very politically involved and entertains large numbers of people on a constant basis. It is designed to be elegant, but also inviting.

The entry hall which was bland and boring is made into what it is by adding Georgian overdoor elements, stenciling the three inches wide planked floors and stenciling and waxing the walls to give them senses of richness and depth. An 18th Irish cabinet and 19th French elliptical mirror endow the entry with magnificent feelings, which sets the tone for the apartment.

设计公司/设计师/摄影师/地点/面积

Design company : Edward Lobrano Interior Design Inc.
Location : New York City, USA
Designer : Edward Lobrano
Area : 743 m²
Photographer : Durston Saylor

本公寓坐落在纽约第79街与公园大道拐角处，它为纽约最富有的家庭之一所有。这是一套家庭公寓，屋主抚育着上一段婚姻的两个孩子，以及他们自己的两个孩子。这对夫妇不仅擅长社交，而且参与政治，并且不断结交大量朋友。所以，这套公寓被设计为一个优雅迷人的家园。

原本乏味无聊的入口大厅加入了乔治亚风格的门头装饰元素，镂花涂装7.62厘米宽的木板假顶，将墙壁进行涂装并打蜡，使其富丽堂皇而又充满深度与内涵。18世纪的爱尔兰餐柜和19世纪的椭圆形法式镜子赋予了入口华美之感，奠定了整套公寓的基调。

客厅：

用琳琅满目来形容本案，我想一点也不为过。满满的艺术气息由地面延伸至墙壁，处处故事处处藏应该是给屋主最好的解释。无论是优雅的沙发座椅，还是精美的墙壁挂画，或者是摆置的各式艺术品，都是关于屋主收藏的故事。这个用爱与时光堆砌的家，静静散发着它特有的光芒！

本案客厅俯瞰公园大道，屋主夫妇酷爱收藏。整个客厅摆设均以其艺术性围绕。其主要艺术收藏品包括毕加索、波特罗、保罗·克利等人的作品。女主人在堪萨斯长大，所以她还收藏了美国当地最重要的画家之一托马斯·哈特·本顿的二十余幅画作。

拥有青色条纹的墙壁、金色的福尔图尼窗帘以及喷漆的木制窗帘框。A＆R阿斯塔大理石壁炉架取代了上个屋主留下的白色木制喷漆壁炉架。3个座位区使客厅有了多种用途，不仅适用于大型派对，当房内只有两人交谈时仍然会有亲密之感。

餐厅：

屋主以前的餐厅完全由布艺打造，所以她希望这里的餐厅采用同样的方式。墙壁采用几乎透明的红色丝绸，窗帘由伦敦的伯纳德·托珀私人订制，路易十五风格的餐椅上包覆着古风皮革和多彩织锦。

SPACE PLANNING ｜ 空间规划

屋主希望设计师首先要考量的是家庭成员的需求，然后是其大量艺术品的收藏，因此设计师在空间规划上以突出此二点为主。以空间功能为主线，艺术品展示为副线，合理布局，凸显家的质感。

儿童房：

小孩房的设计是本案的重点之一，由于屋主小孩多，且都是心头爱，并希望将最好的给予孩子。因此舒适仅变成了基本需求，而艺术与品位则更显重要。每处的设计与摆设，都倍加用心。小房子承载着童年的公主梦，镜面元素是爱美的见证，而小碎花的魅力则更加难以道明。

AN EXALTED HOME
尊贵之家

DESIGN CONCEPT | 设计理念

Evocative of the chateaux of France's Loire River Valley, this estate home, located in Fort Worth, Texas, is designed by the Dallas-based architecture firm, Richard Drummond Davis Architects. Set among expansive gardens and multiple fountains, the home is designed to display the European furniture and collect artistic objects, and it is also a luxurious home created by many treasures. What the furniture and artistic objects give the space is not unordered piles of complicated elements but a delivery of noble, magnificent and gorgeous temperament of the classic French style, which reflects the aesthetic level of the owner and manifests his connotation and taste. Delicacy and particularity are already penetrated into every aspect of everyday life. Artworks are no longer beyond, instead they become essential life tools at hand, which can make the resident live an imperial life.

设计公司/室内设计/景观设计/地点/面积

Design company : Richard Drummond Davis Architects (Architect) Interior design : John Phifer Marrs Landscape design : Lynda Tycher
Location : Fort Worth, TX, USA Area : 1,626 m²

  这套坐落在德克萨斯州沃斯堡市的宅院，是由达拉斯当地的建筑公司——理查德·德拉蒙德·戴维斯建筑公司设计，它唤起了人们对于法国卢瓦尔河谷城堡的回忆。它立于广阔的花园与多个喷泉之间，不仅是屋主用来摆放展览欧洲家具与文物的"收藏室"，同时也是由诸多珍贵藏品打造的一个奢华之家。众多家具、文物赋予空间的不再是繁复元素的无序堆砌，而是一种经典法式的高贵、大方、华丽气质的传达，体现了屋主的审美水平，是屋主涵养与品位的彰显。精致讲究深入骨髓，早已渗入日常生活的方方面面。艺术品不再是高高在上、游离在外的物品，而变成了触手可及、必不可少的生活器具，让居住者享受到帝王般的生活。

SPACE PLANNING ｜ 空间规划

设计公司从屋主一家的生活需求出发，一应空间设置齐全。法式精致、华丽的风格弥漫其中。不管是哥特式风格突出的卧室，还是精致又不失大方的客厅，抑或设计精巧的家庭影院，都为屋主一家深深喜爱。特别是先声夺人的路易十九世风格的圆形大厅和以凡尔赛镜子大厅为灵感设计的走廊，历来被人所称道。

客厅：

设计师采用对称设计的手法，风格偏于庄重大方，整体避开可能暗沉的色彩，颜色多用金色、白色、玫瑰红等色调，令人眼前一亮。细看之下，拥有精美雕刻工艺的钟表摆件，成为艺术品展览在壁炉上。水晶吊灯悬挂其上，晶莹剔透，为空间增添华美气息。橱柜简单大气的外观造型加上绚丽鎏金线条的配置，散发着浓郁的法式风情。增加了花色衬托的地毯，不再古板单一，与玫瑰红座椅相呼应。位于空间视觉中心点的小桌面上暗藏玄机，在灯光的照射下，桌面熠熠生辉，配合桌脚的弧形曲度，显得优雅矜贵。

餐厅：

贵族蓝神秘、优雅，历来是欧洲贵族和王室的代名词。在镶金白色墙壁的衬托下，贵族蓝的色彩显得更加浓烈艳丽，华丽的配色手法极具视觉冲击感。设计师对细节的精心关注与对品质的鉴赏在桌椅设计工艺上得到了体现：镶金边角包裹的餐桌，更显奢华气息；略带曲弧度的椅背，典雅矜持的气质油然而生。无论是就餐，还是在此享受悠闲的下午茶时光，都有一种贵族圈式生活的感叹。白色花卉与空间气质相辅相成，在浪漫优雅的氛围中，顿生清新自然之感。

书房：

　　书房仿造了威尼斯总统府的样式，四周装饰有狮子头的木雕天花板尤为突出。书房四周还陈列着英美作家肖像画，例如马克·吐温、亨利·大卫·梭罗、威廉·莎士比亚和查尔斯·狄更斯等著名作家，旨在学习他们博学多思的探索精神。书房内的家具摒弃了其余空间的镀金装饰和繁复雕花，用平整硬朗的线条追求简洁大方之美的同时，不失古典庄重。天蓝色瓷器摆件与蓝白条纹式样的椅子，柔和的色彩为原本大面积深色调的书房带来轻松愉悦，使人不容易产生视觉疲劳。

家庭影院：

　　这套宅院还包括一间有13个座位的家庭影院，它有别于一般影院的特别之处在于天花板上喷涂的特殊材料，被设计成满天繁星的样式，配合圆形状的穹顶，当光线暗淡时，仿佛星星在眨眼一般，使观影者有种置身于星空下的错觉。

卧室：

　　卧室的气质延续整体风格的金碧辉煌，精致华美不言而喻。床头的精细雕花，庄重大方，典雅气派，充分彰显主人的高贵身份和地位，仿若置身于皇室宫殿之中，回到了十五、六世纪的古典时光。柜子整体样式具有典雅大方的效果，特别是在有镀金装饰的柜边部分，构成室内豪华庄重的气氛。

ENJOYING THE PALACE VILLA
犹享宫廷美墅

DESIGN CONCEPT | 设计理念

This particular project is inspired by the amazing Roman Renaissance palazzo "La farnesina". Symmetry, proportions, architectural detailing and interior design, as well as a luxurious décor, set the tone for one of São Paulo's most celebrated projects.

The owners is a young couple. The wife used to spend vacations in Europe's beautiful castles due to her royal background; the husband is a Lebanese industrial mogul; their two children want to live in a new construction that should appeal to their very refined classical taste. Their home should, nonetheless, be very comfortable and make their guests feel at home either while spending time there or enjoying one of the innumerous impeccable parties hosted by the couple. The classical European style emphasizes the glorious historic sense of the family and pays attention to manifesting the cultural temperament of the couple, which is full of artistic tastes and aesthetic pursuits.

设计公司/设计师/摄影师/地点/面积

Design company : Allan Malouf Studio (São Paulo - Los Angeles)
Photographer : Gabriel Arantes
Designer : Allan Malouf
Location : São Paulo, Brazil
Area : 1,600 m²

  这个特殊项目的灵感来自迷人的罗马文艺复兴宫"法尔内西纳"。对称、比例、建筑细节和内饰设计,以及豪华的装饰,奠定了圣保罗最著名项目之一的基调。

  业主是一对年轻夫妇,妻子因其皇室背景,以前经常在欧洲美丽的城堡度假,丈夫是黎巴嫩工业巨头,另外还有两个孩子,他们想要生活在一个能够体现非凡的古典品位的房屋内。尽管如此,他们的家仍然要非常舒适,无论消磨时光还是享受屋主夫妇举办的无可挑剔的宴会,都要让客人有宾至如归的感觉。古典的欧式宫廷风格营造,在突出强调家庭辉煌历史感的同时,注重业主文化气质的表现,在艺术品位与审美追求间游刃有余。

客厅：

　　双客厅的设置，主次分明，格调统一。除了古典优雅以外，还囊括了大气恢弘。一摆一置，一彩一色，一弧一线都有其内在的讲究。而大小区分有度的艺术收藏挂画的置入，文化气息交织凝练。茂密的欧式古堡丛林壁画，在太阳光的映射中，熠熠生辉。其色彩很好地渲染了厅前的绒布沙发与抱枕，同时金色的边框与抱枕的边角相呼应，协调统一又充满贵族气质。

餐厅：

从房子设计之初的草图到拥有150年历史的巴卡拉水晶吊灯悬挂在正式餐厅天花板上，已经两年了。这个特别房间的墙壁上覆盖着洛伦佐·鲁贝利的真丝锦缎，还有4套木制经典多利安式圆柱与定制的波斯地毯。

门厅地板以四种不同形式的大理石精美地呈现着，与墙壁上娜塔莉·莫朗格的人造石灰岩作品一起，带来了另一个时代的气息。

廊道：

叠见层出的弧形圆拱，创造出室外空间更丰富的层次感。纯洁浪漫的白色线条勾勒出廊道空间的色相，尽头处茂盛的绿植景观，在阳光的照射下惹眼而富有生机，两把休闲藤椅亦可观四方。

SPACE PLANNING ｜ 空间规划

除了可容纳10辆车及1个酒窖的地下停车场外，这座房子有两层楼。在一楼，我们可以看到社会生活区域，如门厅、书房、客厅、餐厅、午餐室（或非正式用餐区），以及1间家庭活动室。在一楼游泳池区域附近，还有1间设施完备的温泉中心。二楼则不同：西侧是1间超大的主卧室，连带其独立衣帽间和卫浴。

为满足家庭需求，在设计公司的总体监督下，打造了一间红木嵌板书房和一间墙壁上挂着莱利公主错视画的非正式餐厅。

书房：

　　每个房间都期待拥有专属的家具，或是来自家庭传承，或是从巴西与欧洲古董商处觅得，书房亦如此。精致的壁炉书柜、案几，精细的挂画、摆件，还有精美的水晶烛台吊灯，一切祥和而美好，充满历史的书卷气质。